D0875717

Beer

Beer

*A Global Journey through
the Past and Present*

JOHN W. ARTHUR

OXFORD
UNIVERSITY PRESS

OXFORD
UNIVERSITY PRESS

Oxford University Press is a department of the University of Oxford. It furthers
the University's objective of excellence in research, scholarship, and education
by publishing worldwide. Oxford is a registered trade mark of Oxford University
Press in the UK and certain other countries.

Published in the United States of America by Oxford University Press
198 Madison Avenue, New York, NY 10016, United States of America.

Library of Congress Cataloging-in-Publication Data
Names: Arthur, John Wood, 1965- author.
Title: Beer : a global journey through the past and present / John W. Arthur.
Description: New York, NY : Oxford University Press, [2022] |
Includes bibliographical references and index.
Identifiers: LCCN 2021035119 (print) | LCCN 2021035120 (ebook) |
ISBN 9780197579800 (hardcover) | ISBN 9780197579824 (epub)
Subjects: LCSH: Beer—History. | Beer—Social aspects.
Classification: LCC TP577 .A7195 2022 (print) | LCC TP577 (ebook) |
DDC 663/.42—dc23/eng/20211109
LC record available at https://lccn.loc.gov/2021035119
LC ebook record available at https://lccn.loc.gov/2021035120

DOI: 10.1093/oso/9780197579800.001.0001

1 3 5 7 9 8 6 4 2

Printed by Sheridan Books, Inc., United States of America

Dedicated to my dad, Ray C. Arthur, and mom, Frances B. Arthur.

Contents

1

Introduction

Beer Is Food

Beer is food for the living and the ancestors; it is not just a cold beverage for a hot day or to have while watching your favorite sports team. I learned this lesson in the highlands of southern Ethiopia, where I lived for two years studying the relationship between pottery and food.[1] My research on beer began not as a conscious decision but through a serendipitous, yet methodical, discovery. As I was studying people's household pottery, I noted each chip and scratch, as well as erosion and soot patterns, and I asked women, "What caused these ks on your pots?" Some of the pots had severe erosion from the lip to the base of the pot. Each woman, when questioned as to the cause of the erosion, stated, "The beer is eating the pots." What does this mean?

I then began to taste Gamo beer, discovering its thick, slightly sour, porridge-like consistency. I participated in gatherings in which women served beer as a way to pay workers for different types of work, such as tilling the soil, harvesting the crops, and building or moving a house. I also witnessed how beer was a symbol of status. Among the Gamo, predominantly wealthy, high-caste families have farmland to produce the surplus +grain to make beer, while low-caste artisans, such as potters and leather workers, generally do not have farmland to grow the grains needed to produce beer. Beer is also a medium between the living and the spiritual world of the Gamo, and all feasts begin by feeding the ancestors first by pouring the beer onto the ground. Apart from the ancestors who are fed beer, beer is also a highly nutritious food that is consumed daily by the living. I began to realize that while people were fed by beer, the process of lactic acid fermenting beer encouraged the beer to "eat" the pots by reducing the pH of the beer.[2] Lactic acid is formed during the fermentation process, creating a delicious sour taste. The importance of beer in Gamo society reveals our cultural diversity as well as our common human potential for creating new technologies for producing nutritious, tasty foods, for socially binding and alienating us with/from each other, for interacting with the spiritual world, and for further motivating us to create, produce, and support one another in work-related activities. I began to wonder if beer was this significant in the lives of other Indigenous peoples[3] in the world and how deep in our history beer exposes our diverse ideas and practices.

In this book, I present a global history of beer as a food produced using malted grains that stimulated innovation in technologies, nutrition and health, and social, spiritual, political, and economic practices. A journey through Southwest Asian, East Asian, African, European, and Central and South American cultures from our archaeological pasts into the ethnographic present reveals the vast tastes, smells, and consistencies of beer that resulted from human technological and social ingenuity.[4] The story of beer originated as far back as 11,000 BCE, when people at Raqefet Cave in Israel began to collect wild grains from which they produced in stone pits a beer— rich in amino acids, vitamins, and minerals and more nutritious than unleavened bread—that was served as a nutritious food that was safer than available water and for propitiating ancestors.[5] Since this time, beer has offered us a path for discovering the diversity in human palates, sophisticated technologies, and complex cultural practices and beliefs. In the past, people created beer recipes with lingon berries,[6] hawthorn fruit and wild grape,[7] and lily,[8] while in the present people use bananas,[9] molle tree berries,[10] and grass seeds.[11]. As a fermented food, beer provides a safe alternative to water and is a source of rich nutrients.[12] Beer was produced in stone pits and pottery vessels, filtered through copper strainers, served in large gourds, and sometimes drunk using long reed straws.[13] The drinking of beer brings together people of similar and different ages, genders, socio-economic backgrounds, and ethnicities but at the same time can be utilized to separate people.[14] The Fur, who live in western Sudan, drink beer from the age of eight months into their old age.[15] In other societies, such as some found in the Amazon and in East Africa, men drink beer only when they are away from their wives.[16] Societies throughout the world conduct rituals in which beer is the central sacrament to the spirit world to ensure community life, fertility, and health. Ancient Andeans poured chicha beer down stone-carved altars as an offering to the sun,[17] and Egyptian tomb workers placed beer vessels in the pharaohs' tombs for them to drink in the afterlife.[18] The Ainu of Japan offered beer to spirits of the mountains as a show of gratitude,[19] and in Zimbabwe beer is presented at sacred shrines located at the site of Great Zimbabwe so that the ancestors will bring rain for a successful harvest.[20] Throughout the world beer is also a motivator that enables people to assist in community projects, such as the Rarámuri of northern Mexico, who use beer as an economic currency of reciprocity to motivate the community to engage in community work parties,[21] or the Pondo people of South Africa, who appreciate work parties that serve beer because they provide a party atmosphere.[22]

The goal of this book is to encapsulate how beer has transformed the economic, ritual, political, and social worlds of past and present societies. By incorporating both the archaeological and ethnographic contexts of brewing beer, my purpose is to demonstrate dramatic cultural change and to give life to the brews by bringing out the actors and actions of brewers and those who participate in drinking beer. Highlighting past and present cultural perspectives is to reveal the variety of ways beer is integrated in people's lives through their technology, health, social status, rituals, and economics.

Technology and Innovations of Beer-Making

Thirst rather than hunger may have been the stimulus behind the origin of small grain agriculture.[23]

Man cannot live on beer alone . . . are we to believe that the foundations of Western civilization were laid by an ill-fed people living in a perpetual state of partial intoxication.[24]

Humans have been hunters and gatherers for 95 percent of the time we have inhabited the earth, and the use of grains to make beer may have stimulated the domestication of grains—one of the most radical changes in our history (i.e., domestication is changing the genetics of plants and animals to the point where they become dependent upon humans to survive). Did beer stimulate the desire to domesticate grains? This was the heart of a debate in the 1950s between Jonathan Sauer and Paul Mangledorf in the quotes above. At that time, there was little to no research investigating the question concerning the role beer may have played in the advent of the domestication of grains. However, some 70 years later the amount of research, both archaeological and ethnographical, has revealed an incredible amount of information about ancient and Indigenous beers. Some scholars assume that early farmers stumbled across the making of beer by leaving grains in a jar with water and realized that this was a way to make a tasty, nutritious, and intoxicating drink.[25] I believe this theory undermines the vast knowledge that our ancestors likely had about the intricacies of plants and how to manipulate them to produce a vast array of culinary treats.

Ground-breaking research indicates that humans have had a long history with eating starchy foods such as wild grains. Researchers have collected bacteria found on human teeth, known as dental plaque, from ancient and modern humans and Neanderthals dating as far back as 100,000 years ago, revealing that humans are adapted to eating starch.[26] This suggests that starchy foods, such as beer, were an important part of our ancestors' diet before the beginning of farming. As we will see, the research on the evolution of dental plaque confirms that beer has played a critical role in the development of our species.

Currently, the earliest evidence for beer dates back to 11,000 BCE in Raqefet Cave, Israel, some 4000 years before the advent of domesticated grains or pottery.[27] These first brewers produced beer in boulder and bedrock mortars using malted wheat and/or barley that was stored in tightly woven baskets.[28] Other evidence of early beer occurred in 7000 BCE at Jiahu, China, where brewers produced sweetened beer made from domesticated rice.[29] In addition, beer may have been produced in Africa before millet or sorghum was domesticated, which may be a strong argument that as beer became more popular there was a need to have better control of grain production, which was the impetus for grain domestication.[30] In the

Americas, wild maize stalks were possibly first utilized to make corn stalk beer, which led to maize domestication and ubiquitous chicha beer that today people enjoy from Mexico to Chile.[31] Brewers created beer some 4,000 years before farmers began to employ domesticated grains in the Old World and possibly in the New World, and once grains were domesticated a variety of grains, including sorghum, millet, barley, wheat, rice, and corn, were brewed. Brewers then added a variety of local ingredients to flavor their beer, among them honey, hawthorn fruit, ginger, garlic, and spicy molle berries, to name only a few.

Containers such as stone mortars and pottery vessels hold the knowledge of past brews. The technological innovation of pottery gave the earliest brewers a place to boil their mash and cool and ferment their beer. The preservation of pottery vessels in the archaeological record has permitted archaeologists to determine that beer was produced in these vessels through the study of use-alteration analyses.[32] Residues taken from the inside of drinking vessels and storage containers in association with funerary contexts have found evidence of early beer production and consumption, such as the fabled King Midas tomb of Gordion in Turkey and the large earthen mound found in Denmark where a high-status woman was buried.[33] It is these lines of evidence in the archaeological record that can be teased out and reflect the social action of beer production and consumption.

Today, Indigenous women throughout the world primarily command the technological knowledge and skills to create non-industrialized beer. They uniquely understand how to refine grain and combine it with other ingredients and retain knowledge of how to produce ceramics and other technologies associated with beer. Beer production requires multiple steps, including harvesting the grain, collecting the water for the malting process (germinated grains), grinding the malt several times, usually on a ground stone, collecting fuelwood or some other type of fuel before roasting the grains and then boiling the mash, and after fermenting for several days, filtering the beer. As I will describe, there are multiple ingredients and recipes that Indigenous brewers use to produce beers that tend to be more bitter and thicker in consistency than industrially produced beers. However, it is the thickness that satisfies people on a daily basis as a food.

Why Indigenous Beer Is Nutritious

Beer has probably been around, in one form or another, since Neolithic times. It has, over millennia, provided clean, uninfected hydration and nutrition for many populations when water supplies and some foods have failed in this respect.[34]

It seems that on Captain Cook's ships beer contributed as many calories to the sailors' diets as biscuits and meat combined.[35]

Today, two billion people only have access to contaminated water, causing an estimated half-million deaths each year.[36] Water-related diseases such as cholera, dysentery, hepatitis A, typhoid, and polio are a major concern for 25 percent of the world's population, and it is estimated that 842,000 people die each year from diarrhea because of unsafe drinking water. The challenge is to relieve this health problem by improving the drinking-water sources of these vulnerable populations by installing piped water sources or protected wells and bore holes.

Meanwhile, Indigenous subsistence farmers provide their own solution to unsafe drinking water by producing low-alcohol fermented beer. While beer is not the cure-all for unsafe drinking water, the fermentation process and the low alcohol content do provide a nutritious food for millions of Indigenous people throughout the world. It is not surprising that alcohol-based drinks became the common form of daily beverages, given the world's poor water conditions, and led to alcohol gaining the title *aqua vitae,* or water of life, during the Middle Ages.[37] The antiseptic nature of alcohol in beer as well as its acidity have the ability to kill bacteria that can make humans extremely ill and cause death. Even in present-day industrial countries, there was a time before the late nineteenth century when water was not safe to consume and the consumption of beer provided a safe alternative to water.[38] The early colonists in America suffered from unsafe water conditions and almost abandoned their posts because of the lack of beer.[39] The Pilgrims who left Plymouth, England, landed in Cape Cod, and because the seamen wanted to keep beer to themselves for the return trip, the Pilgrims were left in their new land without the benefit of their favorite beverage. Fortunately, the water in the area was clean enough that it did not make them sick, but they were ill from scurvy and other diseases.[40] Beer was the one alcoholic drink that the Puritans drank because it sustained good health.

Beer provided a safe alternative to water and also essential nutrition as a food. The consumption of beer in past societies has shown that brewers enhanced the health of an ancient society. In Nubia, present-day Sudan, burials dating to 350–550 CE were found to have the antibiotic tetracycline in their bones caused from consuming beer that contained naturally occurring antibiotics.[41] Today, beer is considered a food adding considerably to the daily calories of many Indigenous societies.[42] Beer is rich in amino acids, protein, peptides, B vitamins, niacin, riboflavin, and phenolic compounds, indicating that drinking beer in moderation will provide health benefits.[43] In African sorghum beers, moderate drinkers consume approximately two liters (half a gallon) of beer a day, which supplies 40 percent of their daily protein and calorie requirements.[44]

Beer as a Communal Act

Beer [is] a drink which is valued, on ordinary days and special days.[45]

Beer must be present in every religious proceeding.[46]

If there's no beer, it's not a ritual.[47]

The importance of and respect for beer as a nutritious food in the daily lives of Indigenous people is highlighted by conventions that beer is predominantly a substance to be consumed socially. Throughout the world, ritual and politics are intertwined in the production and consumption of beer.[48] Robert Netting[49] was one of the first persons to write that beer-drinking among Indigenous societies is done communally, not as an individual behavior. To ensure that these social gatherings are calm and peaceful, there are often rules associated with beer rituals. The quotes above highlight the strong connection between beer and social gatherings, whether they are in everyday or ritual settings. Today, elder Maasai men do not want to lose their status by being drunk in front of young men or women. They would prefer to drink in their houses and would place a stick on top of their house to notify young men and women to avoid the house. One Maasai elder stated, "Alcohol has no respect. It can make you fall in front of your in-laws and sisters and show your nakedness."[50] While beer in moderation does aid in many aspects of people's health, if beer loses its function to socially bind people together and there is a loss of ceremonialism due to social and economic disruptions, then alcohol consumption can lead to serious social and health risks[51]. There is a formality to drinking beer in a communal setting in which the consumption of beer mirrors the social order according to a person's age or sex.[52] The San Francisco Tecospa in the Valley of Mexico teach their children that grown men should be courteous and respectful when they are drinking and that women and children should never become intoxicated.[53]

Ritual participants feed beer to their ancestors, who reciprocate by making the ritual efficacious. Power and rituals can be connected to the dead whereby contemporary Indigenous societies produce and consume beer as a medium that bridges the ancestors and the living.[54] Many feasts around the world are not just to build up the status of particular individuals but are related to important religious components such as feeding ancestors.[55] These ancestors have the power to bring health and prosperity to the living and to appease them when they are fed by the living before any feast begins. In the highlands of southwest Ethiopia, the Gamo catch high-status individuals, who are wealthy and respected by the community, to become ritual leaders who feed the ancestors and community members in beer feasts to bring fertility to all living things.

Serving beer communally also redistributes wealth in order to build political power. In China during the Zhou Period (1027 BCE–221 CE), millet beer was the primary drink for sacrificial feasts in honor of ancestors.[56] By contacting the ancestor, the family member raised the status of the corporate lineage and demonstrated submission to the ancestor as a symbol of parental authority. Later in Chinese society, during the Han Period (206 BCE–220 CE), beer-drinking vessels became a common addition in elite tombs.[57] In the tombs of Meroe, Nubia (Sudan approximately 2,000 years ago), people placed beer-drinking vessels with the dead as a symbol of status by indicating agricultural production in the transformation from grain to beer.[58] Beer serves as an indicator of status and wealth, with tribute payments of beer to leaders when brewers brew a new batch of beer. The Haya of Tanzania are obligated to pay the leader four or more gallons of banana beer in special gourds that are wrapped with banana fiber and tied with twigs and leaves from a plant that symbolizes purity and strength.[59]

Beer helps to bond people together and can foster social communication, interaction, and alliance-building; it can also segregate others from this social bonding by excluding them from taking part in communal drinking.[60] Beer vessels and the formality of serving beer in Canambo, Peru, is associated with daily political diplomacy.[61] The formality of serving chicha also helps to strengthen the social bonds between members of a corporate group (i.e., members, usually a kin group, that owns something together) (e.g., ayllu) organized around kinship and place in highland Peru.[62] The members of the ayllu in Peru drink together to protect themselves from the supernatural because the drinking of chicha and its psychoactive transformation causes groups to be in a dangerous liminal zone.[63] By drinking together, they remain safe from dangerous spirits and channel them to the benefit of the group.[64] The consumption of sorghum beer among the Koma of northern Cameroon by men of different age sets establishes a hierarchy between each one.[65] Upon visiting a household, the Tiriki people of Kenya offer beer to visitors as a sign of friendship,[66] and the Baganda of Uganda use beer as a way to bond two men together.[67] The Tiriki and the Iteso people of Kenya have communal beer pots around which men congregate every day to discuss social issues, settle disputes, and tell stories.[68] The Haya, who live in northwestern Tanzania, pay respect to fathers by offering them a gourd of banana beer, which must be done before others can be served.[69] In addition, the Haya will offer a sacrifice of a gourd of beer at the ancestral altar to a father. These global examples suggest cultural diversity and at the same time reveal beer as of common importance in rituals.

Beer as a Motivating Ingredient

In economic situations, beer is the most widely used recip-rocal for services. . . . The majority of all voluntary labor is repaid in beer.[70]

Beer as a food is used as a commodity throughout the world that can be transferred to encourage labor and social storage, as well as instill prestige and political advantage.[71] Beer used as a commodity has a deep history. Egyptian Pharaohs paid tomb workers with beer, and a document recorded that a taverness in Mesopotamia used beer as a form of payment.[72] During the Wanka I and II periods (ca. 500–1500 CE) in Peru, the local leaders were engaged in the production of large amounts of beer to host feasts to build alliances for work projects and to take part in warfare.[73] Later, during the Inca Period, the state-level production of chicha beer propelled by the Cuzco rulers continued in order to build an even larger alliance based on hospitality.[74]

Beer today continues to serve as a critical commodity to compensate individuals for performing labor for individuals who have the economic means to produce the grains and labor to brew the beer. Subsistence farmers who live on US $1 to $1.50 a day do not have the capital to pay people for their labor in terms of currency, but they do have grains they can transform into beer to pay people to plant, harvest, or build or move a house. The organizer of a work party must provide beer in many Indigenous societies because otherwise it would be impossible to bring people together to cooperate on the task at hand. Beer is used by the Gamo of southern Ethiopia to compensate workers while they engage in community work or by wealthy individuals who can afford to produce enough beer to pay workers.[75] The Kofyar of Nigeria produces beer as the primary means to compensate voluntary labor to prepare and harvest their farms and for building corrals and houses.[76] The Rarámuri, who live in the Sierra Madre in northern Mexico and have been recently popularized by Christopher McDougall's best-selling book *Born to Run*, which describes their prowess in long-distance running, use beer as a payment when they conduct communal work.[77] Beer acts as a binding force among individuals, family members, and

Figure 1.1 Gamo people moving a house, for which they will be compensated in beer.

communities and reinforces the social and economic obligations and reciprocity that cooperative work instills.

What's Brewing Next

We will explore how ancient and contemporary Indigenous beer is intertwined with these four themes: technology, health, ritual, and economics. Indigenous beer adds to people's health, both socially and nutritionally, gives them a reliable commodity to motivate and bond communities together, adds to their spiritual way of life as a medium between the living and their ancestors, and is tied to their technological changes. The diversity of beer production among a number of cultures living around the world both in the past and the present demonstrates a variety of production techniques and ingredients. Just within the past decade, technology has improved to allow us to determine specific ingredients that the ancient ones used to produce their beer. This book takes you on a world tour from Asia, Africa, Europe, and the Americas, discussing the significant impact that beer adds to people's lives, from its role in ritual and communion with the dead to the economic and nutritional improvements it brings to a community.

Chapter 2 surveys the past and present, revealing the diversity of beer-production methods and ingredients. The array of production techniques, technologies, and ingredients indicates that brewers through the ages shared many production methods but also provided their own unique visions of beer. Over time and by region, technologies, gender roles, and ingredients have been constantly changing and in some regions ceased because of the adoption of new religions.

Chapter 3 begins an in-depth geographical journey exploring the relationship between people and beer in southwest Asia, currently considered the region where the world's first brewing took place. Southwest Asia beer documents for the first time new types of technologies that may have been associated with distinct rituals well before the advent of grain domestication. Ancient beer research has now changed our perceptions of when, why, how, and where beer was first brewed. This chapter highlights the earliest beer discovered from Raqefet Cave, Israel, followed with later Southwest Asian sites containing the earliest monumental architecture associated with rituals and decorative stone bowls possibly used for beer production. The chapter discusses the site of Jiahu, where beer first appeared in China and is associated with a rich symbolic context. It outlines the archaeological evidence as to whether beer may have been invented before bread and how beer led to the development of the Mesopotamian state society. The chapter concludes with the ritual use of beer among the contemporary Ainu of Japan.

Chapter 4 focuses on African ancient and contemporary cultures. The chapter begins by examining how African grains were collected and processed before their domestication. After domestication, beers were produced along the Nile as part of the most iconic state societies. New archaeological evidence of Egyptian breweries suggests that the early Egyptian state organized beer production and over time household brewers controlled beer production. Beer's long history in Africa continues today, with beer playing a critical part in today's societies from South Africa to Nigeria to Ethiopia. I will also highlight my own beer research in southern Ethiopia.

Chapter 5 examines the dramatic history beer has played in Europe from the early henges in the United Kingdom to Greek beer production during the Bronze Age to Scandinavia's role in beer production from northern Europe to Iceland. Archaeological evidence from Celtic settlements documents beer-production methods and the connection of beer to the mortuary treatment of their ancestors during the Anglo-Saxon period. Historically, I will discuss the rise of beers through European history and how they have had a major impact on the health, economic growth, and ritual life of Europeans. The end of the chapter will explore the origins and development of hops and European beers from ales to stouts.

Chapter 6 explores beer from northern Mexico to the tip of South America and how the first beers from this region may have come from chewed husks rather than the kernels so ubiquitous in chicha beers. The chapter will look at the development of beers during the different Andean polities, from the ritual site of Chavín de Huántar to the Inca Empire, which intersected beer with fertility, work ethic, and economic reciprocity. I will discuss the rich ethnographic evidence of beers from the Sierra Madres in northern Mexico to the Amazonian rainforest to the montane region of the Andes.

Chapter 7 ends our journey with a discussion of the influence beer has had on our species involving our health, economics, technology, and rituals. The chapter connects Indigenous brewing to the craft beer industry in its efforts toward sustainability and use of seasonal, local ingredients. With the recent growth of craft beer consumption has come a need to educate and train the next generation of brewers, and this has spurred a relationship between craft brewers and universities. An outgrowth of this partnership is creating renditions of ancient beers by brewing arts students working in tandem with local craft brewers as well as experimental archaeologists developing their own brews based on their archaeological research. This chapter includes six of these successful beer recipes, ranging from beers made from stale bread to those infused with chocolate. The art of beer-making extends deep into ancient time, but present-day brewers, working with anthropologists, continue to connect the past to the present.

2

The Diversity of Beer Production

Introduction

Beer often is thought to have only four ingredients: water, grain, hops, and yeast. Yet Indigenous brewers throughout the world over time have produced beers that include a diverse mixture of local ingredients.[1] Brewers enhance the flavor, color, smell, nutrition, strength, and preservation of their brews by selecting water with variety in their mineral contents,[2] diverse strains of yeast,[3] and different types and varieties of grains, as well as other unique ingredients. Societies often take pride in their distinctive beers as a marker of their significant heritage and cultural identity.[4]

The brewing process for ancient and contemporary beers has multiple steps that include malting, milling, mashing, wort separation, wort boiling, straining, cooling, yeast pitching, and the final step, fermentation.[5] Malting is taking the grain, usually barley but it can include wheat, sorghum, rice, rye, corn, and soaking/steeping the grains in water so that they begin to germinate. Usually, Indigenous brewers will steep the grain for a certain period of time and then take the grain out of the water and spread the grains out in the shade so that they can begin to germinate into small sprouts. The sprouts contain enzymes that will convert starch into sugar so that the fermentation process can begin after the yeast is added.

The brewer will then stop the germination process by heating the malted grain, which is called kilning.[6] The kilning process helps to develop the beer's color and flavor. After kilning, the brewer will mill or grind the grain, which is usually done with a ground stone or a mechanized mill. Once the grain is ground up, the brewer will mash the grain by mixing the ground grain with hot water. The wort process starts by first boiling the grain and then separating out the spent grain from the liquid wort. The wort is allowed to cool, and sometimes the brewer will boil and strain the wort a second time.[7] The wort is cooled and usually transferred to a fermenting vessel, where residue yeast adhering to the vessel's interior and/or airborne yeast will inoculate the brew, and the fermentation process, which converts sugars to alcohol and carbon dioxide, will begin. The timeline for brewing ranges depending on the altitude, temperature, and other environmental factors, but usually takes around five days.[8]

The detection of an ancient wine from organic residues extracted from a Neolithic pottery vessel in Iran[9] motivated other researchers to begin the analysis of pottery to determine if beer also could be detected. However, the production of beer can be difficult to decipher in the archaeological record.[10] Residue analysis on the pottery or stone vessels that were used to ferment the beer is the most common and direct method of determining beer production. Other forms of evidence indicating beer production can be brewing areas with specific contextual residues, such as macrobotanical remains (e.g., "malting floors" containing malted grains and strained mash), brewing tools (e.g., ground stones, strainers, and large ceramic vessels with concave lips so that beer is not spilled out of the vessel), and features (e.g., hearths and germination pits) and other use-alteration attributes (e.g., attrition on the interior of ceramic vessels).[11] Iconography, rock art, and texts can be an additional method of inferring production. Residue analysis is the method that determined the earliest evidence of brewing beer occurring in Raqefet Cave at 11,000 BCE,,[12] as well as a number of other sites throughout the world that we will explore in this book.

Since Braidwood and Sauer's[13] debate regarding whether domestication of grains was a result of bread or beer, we have gleaned an incredible amount of information about ancient beers. However, how brewers made their beer and the science behind each decision remains elusive. Beer was being brewed and drunk long before humans invented writing, and even with the advent of writing in the ancient world, brewers did not create a step-by-step manual on how they produced their variety of beers.[14] Currently, there are a number of researchers around the world working to answer questions about how brewers made their ancient beer. Nevertheless, archaeology alone cannot answer these questions. Archaeologists will need to include ethnoarchaeologists and ethnographers into their fold if they wish to learn how Indigenous brewers manipulated their material culture and plants to create the various types of beers. Though current Indigenous brewers make their beer differently from their ancestors, it gives archaeologists a baseline to compare and contrast from what they find in the archaeological record. This chapter describes the global use of technology and plants that past and current Indigenous brewers have used to produce their beers.

Women Are the Brewers of the World

Women were likely the earliest brewers, and we know that in contemporary Indigenous societies women are the primary brewers throughout the world. If we use the world's earliest texts as guideposts, goddesses such as Hathor in Egypt were the inventors of beer,[15] and the Mesopotamian goddess Ninkasi was the beer deity (see later discussion).[16] Beer was not just connected to goddesses; early cuneiform texts state that women brewers also owned the taverns where much of the beer was produced and drunk.[17] Indigenous women brewers continue to brew in many parts of the world, such as Africa, the Americas, and regions of Asia.

In Europe in the eighth century CE, with the introduction of Christianity, there was a movement to establish large monasteries, and men began to brew beer, but household brewing remained the domain of women, with the help of husbands and children, throughout the Middle Ages.[18] In Europe, women who worked in breweries had a range of jobs, from keeping track of what the brewery was producing, to boiling the wort for hops, called *browsters*, or, as a *wringster*, to mixing the malt with hot water. These women had the most difficult job, which is not surprising because women often end up with the most labor-intensive jobs. It was hard work to move the thick dough malt in the mash tun using large and long rakes or paddles.[19] The browster had more status because she was second to the head brewer and she had the important responsibility of making sure that the correct temperatures were reached when boiling the wort.[20] After the plague of 1348–49, beer eventually became industrialized, with an increased standard of living occurring in urban markets, and men

Figure 2.1 Gamo Ethiopian woman making beer.

took over the beer industry.[21] By the fifteenth century, beer made with hops became the beer of choice for most consumers, and since it stored longer it could be transported to more distant locations, encouraging an ever-increased industrialization of beer. In the sixteenth and seventeenth centuries, governments began to regulate beer-brewing by licensing brewers and ale taverns, which promoted monopolization, and implementing taxes. All of these changes over four centuries led to industrialization of beer replacing household brewers with beers produced by male-dominated large corporate breweries.[22]

Today in the United States, men dominate the craft beer industry. Women-owned and co-owned breweries are rare (3 and 17 percent, respectively), and women represent only 29 percent of brewery workers.[23] There are exceptions, such as at America's oldest craft brewery, Yuengling brewery, owned by the four Yuengling sisters in Tampa, Florida.[24] However, in the past and among current Indigenous communities, brewing was and is controlled by women, whereas most of the drinking was and is done by men.[25] Women's knowledge of brewing allows them to gain economic freedom from their husbands, contributing to the education of their children.[26] Although beer production provides additional economic freedom, it is considered by women brewers, such as the Gamo brewers in southern Ethiopia, to be one of the most strenuous and time-consuming activities. Yet Indigenous women today control non-industrial beer in Asia, Africa, and Central and South America, from which industrial beers draw to create more unique recipes.

Southwest Asia and East Asia

Beers in Deep History

From Southwest Asia to East Asia, not only did the continent contain the earliest brewers before the onset of agriculture, pottery production, and a settled way of life, but later beer became one of the main staples for workers building the early state societies. At least 13,000 years ago, hunters and gatherers living in Southwest Asia began to produce and consume beer made of wild grains.[27] This early beer made from wild wheat and/or barley grains was discovered at Raqefet Cave in Israel. They would place the grains in stone mortar-pits lined with grass to ferment their beer adjacent to where they would bury their ancestors.

Brewers in ancient China (11,000 to 9000 BCE) made beer by using two techniques: cereal malts from broomcorn millet and moldy grain and herbs to generate the fermentation process.[28] While using cereal malts is a common global technique for making beer, in Asia the use of mold to instigate fermentation has a long history.[29] Other technological innovations during this time in China include using globular jars for fermenting and storage of beer and funnel steamers for transferring and filtering the wort.[30]

One of the earliest recipes for barley beer also was recovered through analyzing the residues in ceramic vessels in northern China.[31] The brewery at Mijiaya, located in northern China, dates to the late Yangshao Period (3400 to 2900 BCE). Archaeologists found two deep subterranean pits (3.7 and 2.7 meters deep) with wide-mouth ceramic vessels, funnels, and amphorae, all containing "yellowish residues on the interior surface."[32] In addition, a ceramic stove was in each of the pits, most likely used to heat the mash. The type of artifacts suggests that the brewery had three stages of production, from brewing to filtration and storage. Based on the analysis of starch grains and phytoliths (small microscopic grains used to identify plant species), there was a suite of ingredients for the production of beer, including barley, Job's tears, millets, ginger, and lily.

The archaeological evidence for beer production in Southwest Asia does not appear again until 3400 BCE in ancient Mesopotamia, situated in the river valleys of the nation-states Turkey, Syria, Iran, Iraq, and Kuwait. Beer residue recovered from a ceramic jar at Godin Tepe site in the Zagros Mountains of Iran date to the Late Uruk Period 3400 BCE.[33] The large 50-liter jug with a wide mouth has grooves in the interior containing a yellowish, resinous-looking material. Rudolph Michel and his research team[34] revealed that the substance is calcium oxalate, which is produced from processing and storing barley beer, known as "beerstone" by brewers. Calcium oxalate is bitter and potentially poisonous. Potters may have purposely produced the grooves in the pots to collect this compound and keep it out of the brew. Interestingly, the beerstone found at Godin Tepe is identical to the beerstone produced today at the Dock Street Brewery in Philadelphia.[35] Sumerians during this time may have flavored their beers with spices such as cinnamon.[36] Some of the earliest Mesopotamian writing related to beer is from proto-cuneiform text dating to the Late Uruk Period (3200 to 3000 BCE).[37] Mesopotamian brewers brewed nine different types of beer based on the cuneiform symbols of beer jug types.[38] Unfortunately, it is impossible to determine from the text what the nine beer types consisted of.[39] The symbols represent either various ingredients or the size of the beer jugs. During the Late Uruk Period, the texts reveal that barley was most likely the grain of choice for making beer. Scholars have thought that the Mesopotamians were making beer from a type of bread, but a recent deciphered text indicates this as a type of measure for the amount of coarsely ground barley.[40]

One millennium later (c. 2400 BCE), evidence for beer in Southwest Asia appears again during the pre-Sargonic Lagash Period: an ancient text clearly described different types of beer, such as "golden," "dark," "sweet dark," "red," and "strained" beers.[41] Emmer wheat began to be known as one of the leading grains used for beer, but unfortunately the texts do not elucidate how the brewers produced the different types.[42]

One of the most iconic literary texts, the Hymn of Ninkasi, dating to 1800 BCE during the Late Uruk Period, documents early Mesopotamia beer production by women in Iraq. Ninkasi was the goddess of brewing in the Sumerian pantheon. The hymn was accompanied by another poetic text that was most likely a drinking song related to a woman who opened her own tavern.[43] The hymn is one of the earliest texts dedicated to beer production, and every beer lover should pause to appreciate the poem.

> *Ninkasi, you are the one who handles dough (and) ... with a big shovel,*
> *Mixing, in a pit, the bappir with sweet aromatics. Ninkasi, you are the*
> *one who bakes the bappir in the big oven, Puts in order the piles of*
> *hulled grain. Ninkasi, you are the one who waters the earth-covered*
> *malt, The noble dogs guard (it even) from the potentates. Ninkasi, you*
> *are the one who soaks the malt in a jar, The waves rise, the waves fall.*
> *Ninkasi, you are the one who spreads the cooked mash on large reed*
> *mats, Coolness overcomes ... Ninkasi, you are the one who holds with*
> *both hands the great sweetwort, Brewing (it) with honey (and) wine.*
> *Ninkasi, [. . .] [You . . .] the sweetwort to the vessel. The fermenting*
> *vat, which makes a pleasant sound, You place appropriately on (top*
> *of) a large collector vat. Ninkasi, you are the one who pours out the*
> *filtered beer of the collector vat, It is (like) the onrush of the Tigris and*
> *the Euphrates.*[44]

The site of Tall Bazi, located in northern Syria and dating to the Late Bronze Age 1300 BCE, is where archaeologists excavated about 50 houses and found large beer jars in each household that could hold up to 200 liters (50 gallons) of beer.[45] The archaeologists found most of the beer vessels buried under the staircase, where there was good ventilation, and residue of calcium oxalate (beerstone) on the interior of vessels, which is a telltale sign of beer production. The large pots at Tall Bazi are always partially sunk into the floor, which is similar to how contemporary brewers in Gamo Ethiopia place their brew jars. Placing the pots partially into the floor would help to keep the beer cool, and the difference between the temperatures within the pot between the submerged and exposed areas might have helped to create air circulation within the pot that could enhance the fermentation process.[46]

Early Sumerian cylinder seals from the royal cemetery at Ur (c. 2600 BCE) depict individuals drinking beer from straws.[47] Early texts of Enuma Elis, known as the seven tablets of creation and dating to no later than 1200 BCE, refer to the gods drinking beer from straws. Straw-tip strainers have been discovered that date to this time period.[48] Technology changed around 1000 BCE in Jerusalem and Judea, when potters produced ceramic jugs with built-in strainers to filter out the sediment.

Straws helped to keep the unappetizing sediment that floated to the top of the pot from ending up in the drinker's mouth, especially when drinking in poorly lighted conditions. The straws were often made from organic reeds collected along the banks of the Euphrates

Figure 2.2 Beer jug (left) from Jerusalem (tenth century CE) and drinking bowl with handle and sieve (right) from Judea (ninth century CE) (Reuben & Edith Hecht Collection, Haifa University, Israel, Erich Lessing / Art Resource, NY, ART54953). https://www.artres.com/C.aspx?VP3= ViewBox&VBID=2UN365O7HT1DY&VBIDL=&SMLS=1&RW= 1038&RH=601.

or Tigris, but they also produced metal straws and some straws had a metal or bone strainer tip attached to the bottom of the straw to further reduce the amount of sediment that one would ingest.

Later, during the Neo-Babylonian Period of the seventh and sixth centuries BCE, fragmentary texts indicate that the science of brewing was precise in the amount of malt and the timing involved and what was added to the wort to make sure that they could control taste, strength, and color.[49] Mesopotamian brewers brewed large amounts of beer, reaching up to 135,000 liters (almost 36,000 gallons).[50] It seems that the Mesopotamian brewers used a dough called *bappir* that contained malted barley mixed with fragrant herbs that was soaked in water and slowly heated.[51] Brewers then dried the malted barley in the sun or roasted the grains to stop the germination process and pre-serve the maltose sugar.[52] The malt was then ground using a ground stone and sieved to separate the crushed hulls from the sweet malt. The malt was either made into a cake or kept in a ceramic vessel. The texts do not clarify how they introduced the fermentation process, and it may be that the malt was fermented using natural yeasts similar to a Belgian lambic beer (using airborne yeast to inoculate the wort; see Chapter 5). Once the wort was processed, the brewers transferred the wort to another vessel so that the sediment could settle to the bottom of the vessel.[53] Having consistent yeast allows brewers to be certain in their production, and this is why brewers would transport their vessels with them so that the yeast that resided in the bottom and sides of the vessel would help to keep the next brew consistent.[54] The brewers pos-sibly used a vessel that had an opening at the bottom whereby the beer was filtered into a storage and/or transport pot.[55] We are left without knowing how the malting process was stopped, and the mashing pro-cess is only described as "the waves rise, the waves fall," but there is no mention of how they heated the mash.[56] Additionally, we do not know the types of herbs that were used to flavor each beer type.[57] However, the archaeological record can assist in our interpretations of ancient beer production practices. No Indigenous societies today in the Near East or Middle East drink beer because of the widespread prohibition of alcohol in Muslim societies.[58]

Ethnohistoric Beers

In nineteenth- and twentieth-century accounts, there are only a few examples of non-industrial beer production from Asia: the Ainu of northern Japan[59] and from the Tipura region of northeast India.[60] The Ainu of Japan, who are the Indigenous people living on the northern island of Hokkaido, provide a detailed historical account of beer production as part of the house-building rite.[61] After a house has been built, rites are given to protect the house from evil spirits, and part of the rites is the brewing of millet beer. Twenty pounds of millet are ground on a stone and mixed with ten pounds of unground millet. The brewer will take about six pounds of the millet and boil it into a porridge and then pour it into a long shallow trough, allowing it to cool to a warm temperature before yeast is mixed in. The fermentation occurs within a large lacquer tub that is a meter long and deep, and hot embers are placed within the tub and covered with deerskins. Participants will sometimes chew the wort and spit it into the tub to enhance the fermentation process. Once a day for a week, a brewer will open the tub and taste, and on the second day, more mixed porridge and yeast are added with water, and then after another week it is strained. The brewer will strain the beer through a shallow basket, but the beer remains a milky color because the strainer does not completely clear the beer.

In northeastern India, rice is the staple for the majority of alcoholic drinks, but they are produced without malt.[62] However, in specific areas people do germinate rice or millet to make their Indigenous beers, such as the Koyas, who live in the north-central region of India of Madhya Pradesh. This beer, called "landa," has an alcohol content ranging from 2.4 to 4.5 percent.[63] Here, the Koyas will soak the rice for 10 to 15 hours to germinate the grains, and then they are sun-dried in a basket and ground on a ground stone. The mixed powder is then made into small, round cakes, and these cakes are placed on a perforated basket above a ceramic vessel and steamed. Some of these steamed cakes are put in the ceramic vessel with the malt and left to ferment for two to five days. The yeast that inoculates the brew comes from an airborne yeast and/or from the previous brew found in the ceramic pot.

Africa

Beers in Deep History

In the predynastic town of Upper Egypt known as Hierakonpolis (c. 3800–3100 BCE), archaeologists have uncovered evidence for eight to ten breweries, the earliest and most complete material record of brewing in Egypt.[64] A number of breweries have been located in Hierakonpolis with two types of large-scale breweries.[65] The first brewery is a rectangular structure with a low-plastered wall made from stones, large potsherds, and mud and contain rows of ceramic vats. The ceramic vats had a maximum volume capacity of 65 liters (17 gallons). The second type of brewery is semi-subterranean, lined with large rectangular plates, and the ceramic vats were placed in two rows.[66] Each of the eight to ten vats was supported by a series of graduated fire-bars that protected each vat from the heat.

One of the earliest breweries in the world has been located at Hierakonpolis, and in part of the town, designated Operation B, brewers and potters worked side by side.[67] Archaeologists found five

Figure 2.3 Hierkonpolis brewery showing the low-plastered wall with a row of ceramic beer vats.

to six large ceramic vats, most likely used to process beer within a mud structure that contained thick (up to 3 cm), charred residues adhering to the interior of three of the five vessels. The interior residues were dated to 3762–3537 BCE.[68] Chaff and rachis or stems of emmer wheat (cf. *Triticum dicoccum*), with a smaller amount of barley (*Hordeum vulgare* L.) as well as a weed plant, possibly *Lolium* sp. and *Digitaria* sp., were identified from charred material within Vat 4.[69] Besides the residues found adhering to the vessels' interior, charred debris was found in the structure that resembles dried bread containing desiccated plant remains similar to what was found at the ancient Egyptian brewery of Tel el Farkha. The material found at both sites may be discarded wort that was sieved out of the beer.[70] Furthermore, at an adjacent area at Hierakonpolis several emmer wheat grains exhibited morphological features suggesting they were used for malting.[71] It is estimated that brewers at this brewery could have produced more than 300 liters (79 gallons) of beer at a time.[72]

Fermentation of ancient Egyptian beers was were most likely started by using the same vessels to produce the beer, and the residues from previous brews would have provided ample yeasts and lactic acid bacteria to inoculate the wort.[73] Other cultures used the same vessels to begin the fermentation process.[74] Another method brewers may have applied was to add a portion of the last brew to the new wort,[75] similar to what Ethiopians do today when producing the slightly fermented bread injera.[76] They may have left the beer jar open so that airborne yeasts would have inoculated the brew, similar to Belgium's lambic beers.[77] They may have added fruit to jump-start an active yeast growth, but this would be a time-consuming process when so much beer was needed as a staple food.[78] However, it is not clear what technique brewers used to ferment their beers, since they may have used a combination of all of these techniques. However, this area of ancient brewing has not been well studied, leaving a gap in our understanding of how beers may have tasted in the ancient past.[79]

Other than finding archaeological brewing features such as at Hierakonpolis, our knowledge of ancient African beer production occurs in the form of writings, statuettes, clay models, wall reliefs, and paintings from the Old Kingdom (2700–2200 BCE).[80] Evidence of

(a)

(b)

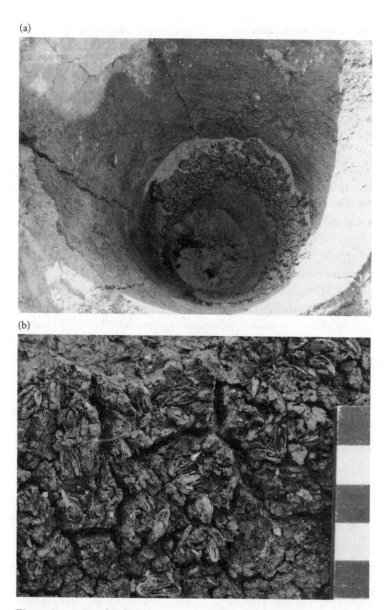

Figure 2.4 a. Residue found in the interior beer vessel from brewing beer at the Hierakonpolis brewery; b. Close-up of the beer residue.

women brewing beer in ancient Egypt comes from a statue of a woman brewer from the Old Kingdom site of Meresankh at Giza.

Other forms of archaeological evidence of brewing are revealed through cereal and starch residues from the artisan sites of Amarna (1353–1336 BCE) and Deir el-Medina (1541–1069 BCE).[81] Evidence of malting using a scanning electron microscope is based on starch granules that are gelatinized, indicating that the cereals were heated while moist.[82] Based on the microscopic analysis of residues, ancient Egyptian brewers were producing beer from bread, with dates as the

Figure 2.5 Figure of an Egyptian woman brewer making beer.

common ingredient.[83] Dates and other ingredients would add flavor and infuse sugar to the mash for fermentation.[84] This is based on tomb paintings and writings that list *bnr*, which is often translated as date fruit;[85] unfortunately, no macrobotanical remains of date fruits associated with beer-making or -drinking have been found in ancient Egypt.[86]

Although different forms of visuals, such as writings, statues, and wall paintings, provide us with a glimpse into the world of brewing, they do not give us the detail we need to understand the nuances of the brewing process.[87] Delwen Samuel[88] conducted extensive studies on more than 200 residue samples from pottery vessels, mostly from sites occupied during the New Kingdom and samples from elite tombs and everyday household contexts.[89] Based on microscopic analysis of the residues, malting was a critical part of the ancient Egyptian brewing process.[90] Lactic acid, which provides a refreshing acidic flavor to beer, also was most likely an ingredient in ancient Egyptian beer, since lactic acid is usually associated with contemporary Indigenous beers that receive spontaneous fermentation from natural yeasts similar to lambic beers today.[91] The structure of the starch granules suggests that Egyptian brewers treated the grains using two different methods. One method was heating the malted grains while moist, since the starch grains were gelatinized such that brewers were stewing the green malt, known as caramalt and crystal malt.[92] Other portions of the malt may have been set aside and allowed to dry, but brewers did not expose the malt to a high temperature. This two-tier system may have allowed ancient Egyptian brewers to have uncooked malt that would have provided active enzymes to break down the starch granules suspended in water, and the cooked malt would have assisted in adding simple sugars for yeast or lactic acid.[93] Residues scraped from ancient Egyptian pots suggest brewers had a diversity of recipes, allowing them to make a variety of beers for people to enjoy.

Details about Egyptian beer are not known, such as how long the malting process took as well as the temperatures, moisture content, stirring rates, and other details.[94] Samuel[95] suggests that the Egyptian brewers took at least three days for the malting process for barley and possibly a day or two longer for emmer wheat, which had a thicker chaff. The malting would have needed a good airflow to prevent the grain from becoming moldy.[96] Brewers may have conducted the malting using large jars turned on their sides.[97] The jars may have allowed for proper evaporation, since the porous clay walls would not have allowed the water to stagnate in the jars and keeping the grains in the jars would have maintained a constant temperature and protected them from other household activities that might have damaged the malt.[98] This type of turned jars with a possible brewer placing his hand into the jar is exhibited in the tomb of Ty at Saqqara.[99] Other side-turned jars are found in two Middle Kingdom tomb models from Beni Hasan and a model from the tomb of Nefery with six jars turned on their sides in front of three brewers.[100] Ancient beer residues are coarse, suggesting that once the brewers finished with the malting process they ground the malt quickly on a ground stone using only a few strokes.[101]

After the quick grounding of the malt, ancient Egyptian brewers mixed the malt with a large amount of water and heated it to a thick porridge consisting of broken grains and shredded chaff.[102] Within the microscopic residues, there are both distorted and undistorted starch granules, suggesting that brewers were inefficient in their processing of grains or that they prepared two separate batches of malted grains, with brewers cooking some of the ground malted grains and leaving some uncooked.[103] Drawings indicate that after mixing the malt, brewers sieved the beer into a large jar, which corresponds with the residues that have only a few fine pieces of chaff and bran.[104] Both the drawings and the residue analysis indicate that Egyptian beer was not made from bread but rather from clumps of damp chaff, and the beer would have looked like modern-day wort, full of starch granules and cereal remains.[105] The alcohol content would have been determined by how much water the brewers added, and texts reveal that ancient Egyptian beers had a variety of strengths regarding their alcohol content.[106]

While texts suggest that Egyptians enjoyed different-flavored beers, there is little evidence suggesting that brewers used additional ingredients to flavor their beers.[107] There would have been a number of ingredients, but none have been identified clearly except for sycamore figs found in a cereal clump taken from a jar from a Deir el-Medina tomb.[108] Paintings show sycamore fig trees lining ritual gardens that were built as shrines for the dead pharaoh.[109] The garden symbolized eternal life, and figs were associated with the goddess Hathor.[110] Text images in the Book of the Dead show Hathor living in the syca- more tree, where she was referred to as the "Mistress of the Southern Sycamore" who gave out food and water and protected the dead souls from being burned by fire. It should be noted that the coffin of Osiris was made from the sycamore tree, giving him eternal life.[111]

The British Museum conducted an experimental research project to re=create ancient Egyptian beer using archaeological reports and chemical analysis of pots, as well as input from curators and phys- ical anthropologists to help guide them in producing a beer as sim- ilar as possible to the ancient Egyptian brew.[112] They used emmer wheat, which is the ancient form of wheat and, as discussed earlier, has been identified in the context of ancient Egyptian breweries found at Hierakonpolis and Amarna.[113] To add some spice to the brew, the ex- perimental archaeologists added rose petals, pistachios, sesame seeds, coriander, cumin seeds, and dates. Using research methods determined by Samuel,[114] they used a two-stage mash, whereby the cold mash uses ambient-temperature water and malted, ground grain so that the mash would convert the starch to sugar. The second mash also contained malted, ground grain, but it was mixed with hot water and further heated to about 80 degrees Celsius, and then eventually the cold and hot mashes were mixed together. The final mash was left to cool to turn the starches into sugars, and then it was sieved and placed in a ceramic vessel that had been inoculated with yeast. A potter re-created the an- cient Egyptian pot with a wide, open mouth to help expose the brew to natural yeast, left the pot unglazed, and used highly porous clay to naturally cool the fermented brew. The mouth of the ceramic vessel was covered with a muslin cloth, and the fermentation process began. The brew would have been most likely drunk using either a ceramic or reed straw with holes at the end constructed as an internal strainer to keep unwanted particles from being drunk. However, the brew that the experimenters produced was not thick but had the consistency and color of a modern lager or ale. The final brew ended with a description by the researchers: "it was absolutely delicious!"[115] All of the archaeo- logical research on ancient beer production comes from the Nile Valley, but ethnographically we have a far richer documentation of contempo- rary beer- production throughout sub-Saharan Africa.

Ethnographic Beers

The production of Indigenous beer throughout sub-Saharan Africa is part of the culinary traditions created by women, who produce beer using different methods, styles, local ingredients, and flavorings. The biogeographical diversity of Africa allows for a range of ingredients, such as bananas and millet in the East African lake region of Tanzania, to barley and wheat in the highlands of Ethiopia, to sorghum beers in the lowlands regions of Sudan, West and East Africa, and South Africa. Now, with the introduction of corn in the past 500 years, corn beers have grown in popularity across large portions of sub-Saharan Africa. The technology of brewing includes using ground stones or a mechanized mill for grinding the grain, fermenting the brew in large ceramic vessels or gourds, and straining the beer using woven grass.[116] African beers use spontaneous fermentation of the wort that may result in a varied taste and aroma from brew to brew.[117] African beers, as with other Indigenous beers throughout the world, do not have a long shelf life, usually lasting from three to five days before the beer spoils. The consumption of beer also varies, with regions using gourds for drinking and other areas imbibing their brews communally with long reed straws. These beers also reflect ethnic identity wrapped in symbolic and economic agency, which is further discussed in Chapter 4.

African women impart an incredible amount of skill and knowledge to producing a diversity of beers that requires a considerable amount of time. Beer production among the Mabaso women of South Africa, for example, requires 10 to 12 days.[118] Brewing time for the Maale of southern Ethiopia takes five days of extremely difficult labor that includes grinding the grain for hours, collecting water from at least a mile away, gathering the fuelwood to boil the malt and flour in large pots, and then fermenting the wort in large gourds.[119] The Pondo women of South Africa bring their own grindstone to a specific household and grind grain together long into the night to produce enough beer for a feast, taking three to four days for the entire brewing process.[120]

The Gamo, who live in the highlands of southwest Ethiopia, produce beer using a range of tools from ground stones, different ceramic vessel types, and a grass-woven sieve.[121] The production of beer is associated with the wealthy and high-caste Gamo households. They use different types of grains depending on their ecological zone, which determines whether they subsist more on barley, wheat, sorghum, finger millet, or corn. They sometimes add garlic, ginger, or pepper for flavor.[122] The Gamo germinate their beer by placing about two to three kilos of grain in a large ceramic bowl with water and covering the bowl with enset leaf. They allow the grain to germinate in three days, and the brewers dry the grain outside and then grind the malt on a ground stone. After grinding, they heat the grain on a large ceramic baking plate, and before the grain burns, they add the roasted malted flour and water to a large ceramic jar. After a day, the wort begins to ferment, and they eventually will sieve the wort through a grass-woven sieve into another, larger ceramic jar. The Gamo drink their beer using gourds that have been decorated with designs usually burned into the skin of the gourd.

In Burkina Faso, the Maane brew beer for both ritual and commercial consumption, both under the domain of women.[123] Beer that is brewed with red sorghum symbolizes the act of creation, similar to a baby being born. The production of beer brings the ancestors to the house during key times of the year, such as the growing season of the red sorghum. The ancestors who reside in the west return to the east to join the living in line with the movement of the first moon and the next 14 days, when the moon walks as do the ancestors from the western sky to the east,.[124] whereas the funerary rituals occur years after the person has died and is in step with the 14 days of the waning moon. The sorghum harvest has three monthly rituals paralleling the pregnancy and eventual opening of the eyes of the sorghum. Each ritual in this three-month cycle coincides with the brewing of beer and is orchestrated with the chief leaving his palace so that the ancestor chiefs may arrive from the west. During this ritual, the senior women collect ripening ears of the red sorghum and place them in the pot when the beer begins to boil. This represents the sorghum opening its eyes, and it is born just as the beer is.

The red sorghum is sprouted by wetting the grains, and after this process, the context of where the consumption is going to take place dictates who and how the beer is brewed.[125] If the beer is produced for ritual purposes such as a funeral, then the senior wives are solely responsible for grinding the grain, and they should speak loudly as they grind the grain. Younger women are not allowed to grind grain for beer that will be consumed in a ritual place. Large amounts of water are gathered and transported by women back to the household. This underscores the incredible amount of labor that goes into collecting water, sometimes several miles away, to produce beer as well as for daily use. In the household courtyard, the malt is mixed with water, and during the lautering process (i.e., separating the liquid sweet wort from the solid spent grain) when the heaviest malt has sunk to the bottom of the pot, the surface water is poured into large beer jars located on the hearth and heated for an hour or so. If the beer is produced for a ritual, then the beer water is fed to the ancestors before being placed on the hearth. After what the Maane call the cooking, the wort is cooled for the process of souring, which is when the lactobacillus and pediococcus bacteria grow and contribute to the sour flavor.[126] The best brewers know exactly when to stop this brewing step, which may occur in the middle of the night. The heaviest fraction settles to the bottom of the pot, and then the surface water is heated as well as the sediment, which is filtered and remixed with the malted beer.[127] Yeast is then added to the beer, which is given the name "baby" and allowed to ferment for a day and a half before it is ready to drink, maturing to the name of "old person."[128] The adding of yeast is associated with a baby crying, and a brewer may ask one of her children who is known to cry to add the yeast to get things off to an auspicious start.

African beers are made with a variety of flavoring ingredients (see Table A.2), but some areas produce beer with the primary ingredient being banana mixed with sorghum.[129] In Tanzania, the Haya make a beer using a combination of bananas and sorghum and having a low alcohol content of about 4.5 percent.[130] They use ripe bananas that have been hung near the hearth or buried in a pit for several days. The ripened bananas assist by increasing the sugar content, which will promote the fermentation process. The bananas are smashed with the help of dried grass in a wooden trough, and water is added in equal proportion to the banana liquid. After the water is added, dried sorghum is mixed into the water and banana solution, and the trough is covered with banana leaves and left to ferment for 24 hours. The mixture is never heated. Women and children can help with collecting certain materials, such as the grass or water, but women are not allowed to smash the bananas and men, who must be sexually abstinent during the brewing process, do the majority of the work. The consumption of the banana beer is gender-specific, with men drinking from gourds with long necks and sipping the beer from a hollow reed while women drink from short-necked gourds without a straw.

Europe

Beers of the Deep Past

The vast variety of beer styles and their associated regions in Europe today are considerably different from those of the ancient European past.[131] The earliest evidence of Nordic brew comes from the site of Nandrup, Denmark, at around 1500 BCE, where an earthen tumulus tomb was found on the island of Mors in northwest Denmark.[132] Nordic groups extended their trade with the rest of Iron Age Europe and the Roman Empire, which brought them into contact different traditions of ancient alcohol .

Beer-brewing was a household activity with varied brewing traditions between households and regions, with the local ingredients and water quality giving each region its own unique style. Eventually, brewing became a more specialized occupation. For example, in Britain we find the maltster Optatus and brewer Altrectus, who worked for the Roman army of Vindolanda around 900 BCE.[133] Other regions developed specialized brewer guilds in the first centuries CE along the Mosel River, which runs through Germany, Luxembourg, and France today, and in Trier, Germany.[134] By the ninth century, monastic breweries had become well established. Unfortunately, before 1000 CE there is no evidence of how they were brewing their beers and what styles they were making.[135] Even when there are descriptions of beer, the details are vague, with only a mention of grain type or additional ingredients but without any specifics.[136] For example, during the ninth century CE in Ireland, descriptions of brewing state that the malting process of barley took between 12 and15 days. Brewers steeped the grain for 24 hours and drained for 36 hours, four and a half days protected under cover and then three days exposed in piles, and raked into ridges, before being finally dried in a kiln.[137] Furthermore, the archaeological evidence can be ambiguous as well, with malted grains indicating either beer production or bread-making. However, there are archaeological sites where it is inferred that beer-malting was occurring, such as at the Iron Age site of Eberdingen-Hochdorf in southwest Germany and dating between 600 and 400 BCE.[138] Brewers dried the malted grain on a series of wooden lattices placed over fire pits. Archaeological evidence found charred barley grains in six U-shaped ditches that were five to six meters in length, 60 centimeters wide, and one meter in depth. Wooden boards supported the edge of the ditches, and the brewers lit fire pits at the end of the ditches to gently dry the malted grain. However, the fire grew out of control and burned the malted grains in the trench, preserving the remains for archaeologists to find two millennia later.

The type of grains used in ancient European beers primarily consisted of barley, then spelt and bread wheat as secondary grains, and millet beer brewed in eastern Europe.[139] Rye beer was common in Estonia up until the nineteenth century.[140] Other types of grains used by ancient European brewers were emmer wheat and oats, and brewers in Colchester, England, may have used malted spelt wheat and barley together, since these macrobotanical remains were found with a ratio of 10 to 1.[141] It is possible that spelt wheat was the preferred grain in southern Britain, whereas in the north, barley was the predominant grain brewers used.

Archaeobotanical remains suggest that brewers were using a range of additives, especially during the Middle Ages.[142] Two of the most common ingredients added to beer were sweet gale or bog myrtle (*Myrica gale* L.) and hops (*Humulus lupulus* L.). The shrub sweet gale grows in northwest Europe in acidic bogs and sandy soils and contains a fragrant scent from the gale oil that is excreted though the glands in the leaves and flowers.[143] It was most likely in use as an additive for beer during the pre-Roman Iron Age based on the ubiquity of the archaeological sites with evidence of sweet gale in good context.[144] These sites cluster around the present-day Netherlands, and then in the Middle Ages the dispersal of sweet gale for brewing is found in natural habitat areas but spread into what is now Scandinavia, Britain, and Germany. While sweet gale is not used today, based on early texts, it is argued that apart from malted grains it was the primary ingredient in these early medieval beers up until the thirteenth century.[145] The earliest mention of sweet gale is in Latin, *materiam cervisiae*, from 974, CE when Emperor Otto II transferred the sweet gale rights of Belgium to the church at Liege. Later, the sweet gale is referred to as gruit or grut in the text, and the first time it is mentioned dates to 999 CE, when Emperor Otto III mentions grut as a trade item to the church of Utrecht.[146] By the fourteenth century CE, beer made from sweet gale began to decline, especially after rumors surfaced that sweet gale causes blindness and eventual death.[147] In the Netherlands, by the sixteenth century CE there is only slight evidence that grut beer was still being brewed.[148] Chemical analysis does not indicate that sweet gale is dangerous to ingest, but based on these false perceptions, the production of *Myrica* beer was forbidden in northwest Germany by the early eighteenth century CE.

Hops occur naturally in the temperate regions of Europe and extend into Scandinavia and the Mediterranean region.[149] Brewers use the hop's female cone that contains bittering agents lupulone and humulone, which help to flavor and preserve the beer. When only a few hop samples are found, it is difficult to assess if brewers were using them in their brews, because hops can grow naturally around ancient settlements. However, sites that have good context with large amounts of recovered hops provide a stronger inference that hops were part of the beer recipes for past brewers.[150] There are too few early sites to determine where brewers may have started adding hops to their beers.[151] The oldest evidence of hops dates to the early medieval period from the Develier site in eastern France (sixth to eighth centuries CE),[152] the Serris-Les Ruelles site in central France (seventh to the ninth centuries CE),[153] and the Mikulcice site in the present-day Czech Republic (seventh to ninth centuries CE).[154] The oldest text reference to hops occurs from 768 CE, when king Pipin, father of Charlemagne, donated hops to the monastery of St. Denis in Paris.[155] However, it was not grown by the Charlemagne estate at this time, since all plants were recorded by the royal estate, but it was most likely collected at this time and place. The first record of hop gardens is from 859 to 875 CE from the documents recorded at the Hochstift monastery in Freising, Bavaria, in Germany. The Graveney boat site located in Kent in southeast Britain, dating to the tenth century CE, strongly suggests that hops were being traded for brewing purposes at this time.[156] Hops provided bitterness to the beer and was an important preservative. Before hops, beer did not have a long shelf life and had to be drunk most likely within a week's time, similar to most Indigenous beers produced around the world today. Hopped beer was more prevalent in areas where sweet gale did not occur naturally, and over time hop beer was traded from Lubeck, Wismar, or Danzig, Germany, to Denmark and Sweden in large quantities by the thirteenth century and then expanded into more areas after this time.[157] Brewers over time learned that boiling the wort for approximately three hours was optimal for extracting hops' preservative and flavor qualities.[158] However, brewers sometimes boiled the wort for 20 to 30 hours to achieve a higher alcohol content. Today, Bavaria and the Czech Republic are the dominant hop-growing regions in Europe.[159]

Other ingredients were added to European beers, based on texts recorded after the 1500s.[160] This is surprising, since the Bavarian Purity Law (Reinheitsgebot) was instituted in 1516 CE. Bavarian beer law restricts brewers from using any ingredients other than water, yeast, hops, and barley malt for bottom-fermented beers, and for top-fermented beers (see glossary), brewers must use the same ingredients but can use wheat malt as well.[161] Although the purity law allows for only four basic ingredients, German brewers were able to produce a variety of beers from bocks to lagers to pilsners. Dried ingredients of stems, roots, leaves, and flowers as well as sweet fruits (i.e., cherries, blackthorn/sloe berries, raspberries), honey, and sugar were added for taste rather than as a preservative, since these brewers were using hops as well.[162] Some of the ingredients were being used until recently, such as marsh Labrador tea (*Ledum palustre*) and juniper (*Juniperus communis*) in Estonia.[163] Other ingredients, such as marrubium mint, honey, and a host of other ingredients, were included in special beers, usually ales, to treat diseases such as the plague and lung disease.[164]

Honey was fermented with malted grains in ancient Europe to produce a honey beer or bragget.[165] Adding honey to beer brought many benefits, with honey increasing the alcohol strength, helping to preserve the beer; providing yeast to start the fermentation of the malted grain, giving a sweet flavor; and possibly adding a narcotic feeling to the beer from the flowers and nectar.[166] Archaeological evidence from residue analysis confirmed the existence of honey beer at the Lichterfelde site near Berlin, Germany, dating to 1000 BCE, and -a Celtic grave in Glauberg, Germany, contained honey beer.[167] Textual evidence of honey beer exists from medieval Britain, especially from Wales, where the Welsh laws written in the tenth century CE state that the king should be provided with a vat of mead large enough for him to bathe in, or two vats of bragget or four vats of beer.[168] Malted wheat was likely the grain type mixed with honey to make the bragget, giving it a cloudy appearance, in contrast to the clear ale possibly made from malted barley.[169] Brewers stopped making honey beer during the Middle Ages, but British and other European brewers revived this brew type during the twentieth century.[170]

> *Sixteenth-century brewers were certainly not aware that there were some 350 species of yeast.*[171]

European brewers borne yeast to inoculate their brew, and adding yeast did not occur from the beginning of beer production through the Middle Ages.[172] The airborne yeast would affect the hot wort after boiling, or brewers could add some beer from the previous batch, or add bread to the beer, or not clean the fermenting troughs very well so that the next batch would be infected with the previous brew.[173] The methods of using yeast for European brewers changed through time, especially during the fifteenth and sixteenth centuries.[174] Eventually, brewers realized there were two types of yeast, a yeast that rises to the top of the wort and another that settles to the bottom of the fermenting vessel.[175] It is possible that brewers regulated their yeast strains, which they kept separate.[176] Pure yeasts did not occur until the end of the nineteenth century, but brewers were most likely skimming the foam off the fermented beer to start the next batch.

Seasonal temperature changes affected the types of yeast brewers would use, with bottom yeasts needing colder temperatures, between six and eight degrees Celsius, and brewers using top yeasts in the summer.[177] Places such as the Netherlands and Belgium, with a milder climate, made using bottom yeasts more difficult, so brewers would begin their work early in the day, before dawn so that they could put the wort into the fermentation troughs in the cool of the evening. The troughs were purposely placed so that cool air would flow over them, and eventually brewers would use fans to push the cool air across the top of the beer trough. Summer was a time that yeast could be destroyed, so brewers would dry the dregs from the beer barrel and mix the yeast with flour to form cakes. When water was added, the yeast would begin to grow.[178] Or the dregs would be kept wet and then added to the next brew to start the fermentation process. In the sixteenth century CE, bottom fermentation took 10 to 12 days, whereas top fermentation was much faster, taking only one to three days. To reduce spoilage, the brewer wanted to minimize the amount of air that the beer was exposed to, and brewers using casks would place rough paper in the bungholes before tapping in the bungs. Brewers would also use deep fermenting troughs to reduce the amount of beer exposed to the air and use larger and deeper troughs made for better beer.

Brewers were regulated by governments as to how long the beer had to stay in the brewery. In 1392 Amsterdam breweries were required to keep beer in the brewery for four days before it could be sold to consumers, but beer that was sold outside of town could be shipped the same day.[179] Later regulations, known as the Delft bylaw on beer in the Netherlands dating to the fourteenth century, mandated that the brewer wait eight days from the time the beer was put in the barrel before it could be transported to North Holland. However, as with the earlier regulations, overseas beer was not regulated in regard to when it could be shipped out. Brewers would add herbs and eggs and sometimes linseed oil to help preserve the beer and improve the taste.

Ethnohistoric Beers

The tools European brewers used from 1450 to 1620 CE changed at a slow rate.[180] The amount of technical writing about what brewers did was limited, but what information we can gleam from the writing by brewers was in response to governmental regulation and taxes. Records indicate that the type of materials brewers used to start a brewery required a considerable amount of capital, usually done by wealthy families, such as a tun for mashing, a kettle for boiling the wort, a cooling trough, a fermenting trough, and barrels, as well as shovels for stoking and moving the grain, and rakes and paddles for stirring.[181] An Amsterdam brewery that was sold in 1511 lists the type of equipment,:

> two fires, two kettles one of about 25 and other of about 35 barrels or approximately 3000 and 42000 litres respectively, a mash tun, a yeasting tun of about 70 barrels or some 84000 litres, three troughs used to carry the beer to the three cooling vats of some 7.25 meters by 3.9 meters or the small cooling vat of about half that size. There were 13 racks, presumably for storing barrels . . . There was a maltery as well with an apparatus, made from stone and cement, for storing and sprinkling water in the large malt attic, There were two grain attics, a kiln for roasting which was equipped with iron latticing and additional equipment. The building included a peat rack for storage, a loading stage at the side of the house and a flat-bottomed inland boat complete with masts and ropes for shipping water to the brewery.[182]

By the fifteenth and early sixteenth centuries, the quality of the kettles, troughs, and other implements had improved. Other improvements occurred, including placing the iron or copper kettles over an iron grate with walls on top of a furnace. Brewers constructed a brick wall surrounding the furnace to allow workers to more easily stir the work, which had the added benefit of saving fuel and reducing the amount of smoke from the fires. By this time, all urban breweries had brick ovens with large kettles and plumbing that transported water and wort to and from the kettles.[183] Copper kettles were considered the best type of kettle in the fifteenth century and they were the most common. Wood continued to be used for troughs and vats, and by the eighteenth century, brewers began recommending "hard and dry oak which had been cut across the grain for fermenting troughs."[184] Brewers used straw in the vat to strain the wort, but in 1501 CE they received permission from the Dutch government to change the technology of the mash tun using a false bottom in the tun to keep the spent grains from the wort, providing an easier way to obtain the wort.[185] Otherwise, brewers bailed out the wort using bowls or ladles. Brewers used a number of different options during the boiling of the wort to make their beer more clear, such as pig or ox foot, burned salt, clean sand, lime, ground oak bark, and dried fish membranes.[186]

Americas

Beers of the Deep Past

The brewing of beer in Central and South America with corn/maize and other associated ingredients has a long history.[187] Evidence of beer production has not been found in North America north of the present-day United States-Mexico borderlands. Botanists and archaeologists have theorized that the earliest brewers in the Americas were making beer not from the maize kernels but from the maize stalks.[188] It is argued that the impetus to domesticate maize was for the sugary stalks and not for the grains and that the Indigenous people who occupied the caves of the Tehuácan Valley in central Mexico chewed the maize stalks to produce a sugary drink. While excavating a series of caves in the Tehuácan Valley, archaeologists found 83 chewed stalks or leaves and 140 chewed husks dating from 5000 to 1540 BCE.[189] Based on starch residues from ground stones found in a Xihuatoxtla rock shelter in Guerrero, Mexico, domestication of maize occurred at 6750 BCE.[190] In Bolivia, Forbes[191] reported in the late nineteenth century that young sweet stalks were fermented into a drink, but today they are only in the markets to be chewed, like sugarcane, and they are no longer brewed into a chicha beer.[192] In Mexico, the Huichol Indians also make a fermented drink from corn stalks.[193] Thus, there is ethnographic evidence that corn stalks could be processed to make an alcoholic drink, but beer from kernels is the most common form of a grain-type beer in Meso- and South America. This new evidence does not negate the sugary stalk hypothesis, but it is possible that brewers were making maize beer from the kernels as well as from stalks.

Brewers in the Americas used ceramic vessels to produce maize beers, and as demand increased over time, the size of the vessels increased such that professional brewers today in the Andes produce beers using vessels that can hold up to 170 liters (45 gallons) of beer.[194] In the past, Andean brewers also used large pots and have been found in the context of large breweries associated with the earliest state societies.[195] One of the best examples is from the site of Cerro Baúl, dating to the Wari Period (600–1000 CE), where archaeologists uncovered the largest pre-Inca brewery with 12 150-liter fermentation jars that could produce 1800 liters of beer.[196] Later, during the Inca Period (c.1400 to 1533 CE), they used similar-size ceramic vessels to produce thousands of liters of beer for Inca feasts.[197]

New technologies can indicate changing cultural identities, such as the production of the *kero*, a tall and intricately made drinking cup, produced by potters during the reign of the Tiwanaku polity, which was a state-level society that began around 500 CE and lasted to about 1100 CE.[198] The *kero* is associated with drinking chicha beer and is found in many different social contexts, suggesting that individuals drank chicha from the *kero* at all social strata. In fact, the *kero* drinking cup was the symbol for the Tiwanaku state, with its beautifully crafted iconography, burnishing, and thin walls.[199] At Cerro Baúl, 28 *keros* that ranged in size from 12 to 64 ounces were thrown into the brewery at the ceremonial burning of the site (see Chapter 6 for more discussion of Cerro Baúl).[200]

Chicha beer production is the domain of women throughout the Americas and has been since at least the fifteenth century CE.[201] The preparation of maize beer today is similar to production steps brewers used from at least the Early Intermediate Period (200 BCE to 250 CE) and continued into the early twentieth centuries; however, this is not to suggest that there were technological, sociocultural, and economic changes that affected beer production over time.[202] Men and children do help with some of the early tasks of the brewing process, such as harvesting the maize or other ingredients and collecting fuelwood, but women control the production process throughout.[203] Beer is brewed within the household area, and the types of technology brewers use are ground stones for grinding the grain and large pottery vessels (ranging from 40 to 120 liters) for boiling the wort and the final fermentation process.[204] The alcohol content can range from 2 percent for new or watered chicha to as high as 12 percent; however, most chicha beers are around 5 percent alcohol or less.[205]

Ethnographic/Indigenous Beers

Ethnographically, and most likely archaeologically as well, there is variation in how chicha beer is produced Throughout the Andes region;.[206] There is diversity in the types and colors of maize brewers use to make their chicha beer.[207] Differences range from the degree of milling of the corn, sieving practices, the amount of sugar added to increase fermentation and alcohol content, fuel wood types, cooking times, and length of fermentation. For example, based on historical descriptions, Andean brewers made beer ranging from two types of red beers to a yellow colored beer and a clear beer.[208] Historically as well as today, northern coast Peruvian brewers sieve their beers twice, using a coarse and a fine sieve.[209]

Brewers in the Bolivian highlands make their household beer by either salivating the corn flour balls or letting the maize kernels germinate for 12 to 18 hours and then grinding the sprouted kernels into a malt on a ground stone.[210] The flour balls are worked with the tongue until the balls are well moistened with saliva.[211] The salivated balls are then dried in the sun and stored or transported in sacks. Beer made from salivated balls is in higher demand than unsalivated malt, so brewers try to convert as much flour as possible into the salivated balls. In the sixteenth century CE, there were descriptions from the southern Peruvian coast where specialists were paid to chew toasted maize flour.[212]

The malted maize kernels are soaked overnight in a pottery jar with water that completely covers the kernels.[213] The following day, the kernels are spread out to about two to four inches thick on leaves and covered with several blankets.[214] The blankets cause temperatures to rise to as much as 34°C, and when the shoots are as long as the kernels, they are placed in the sun to dry ranging from two to five days, allowing for an easier time to either grind or mill the maize.[215] After the maize kernels dry, they are ground on a ground stone.

The ceramic vessel is filled to approximately one-third with dried and ground malt or salivated flour.[216] Brewers will sometimes mix unsalivated flour with the salivated balls as well as sugar or the pulp of a squash. The large open-mouth ceramic vessel can range in volume from 40 to 120 liters for heating the wort to about 75°C.[217] The first boil can last up to three and a half hours, and brewers will add water to the vessel, since evaporation occurs and brewers remove sediment and the sugary upper layer.[218] A second boil can occur, lasting about four hours, and then the wort is allowed to cool and clear. On the fourth day, the wort begins to bubble violently and ferment, and the beer is sieved either through a cotton cloth or wire sieve into narrow-mouth ceramic vessels in which it will be consumed or sold.[219] Brewers sometimes will use a gourd dipper to transfer the beer between the two vessels rather than moving the pots, especially if the pots are extremely large.[220] Fermentation can take up to six days, and if the beer is produced at a higher altitude and cooler temperature, the fermentation process can take longer, whereas in the lowlands fermentation can take as little as two days.[221] Before the chicha is consumed, the top layer of oily, yeasty froth is removed with a cupped hand and is used to polish wooden furniture or added as a starter to a new batch of chicha. However, because the pottery vessels are never scrubbed but just lightly washed, there is enough yeast to activate the fermentation for the next chicha brew. The layer on the bottom of the fermentation vessel is taken out (*sutu*) and laid on a gunnysack stretched over a small jar. The liquid that drains into the jar is prized for its higher alcoholic content compared to chicha and is sweetened with sugar and dyed with prickly pear cactus seeds.[222] About 40 liters of chicha will produce approximately one liter of *sutu*. The remaining dry part, called *borra*, is fed to pigs and chickens, made into a low-alcohol chicha, or added to bread dough. About 100 pounds of shelled maize will produce 14 to 15 gallons of chicha beer.[223]

In the lowlands of Bolivia, the dull orange maize is the common type for making chicha and, as in the highlands, the kernels are soaked in water, then placed in a basket lined with Heliconia leaves.[224] This raises the temperature in the basket to 36°C, and when the shoots reach the same length as the kernels, they are ground with a ground stone. The brewers then salivate some of the malted flour, and all the malt is added to a ceramic jar and mixed with warm water. The wort is boiled for three to four hours, left to cool, and eventually strained through a piece of cotton cloth for fermenting to begin. The chicha is most often drunk immediately after it is cooled and strained, but it is best after two days.

Because of the warm lowland temperatures, the beer does not keep more than five days.

Other regions in the Americas that make chicha are in Nicaragua where the Nicarao and Chorotega Indigenous communities make a beer from ground corn and honey.[225] As well, the Huichol Indians, known for their long-distance collection of peyote in northern Mexico, also make a chicha corn beer.[226] After the Spanish arrived in the American Southwest, they introduced wheat to the Native Americans, and the Yuma began making a wheat beer.[227] As far as the archaeological and ethnographic research indicates, this is most likely the northern-most point in the Americas for Indigenous beer production.

Various other chicha drinks are made throughout the Andes that do not include maize, such as one from a species of *Mauritia* palm[228] made in the Bolivian lowlands as well as molle chicha made from the *Schinus molle* tree berry found mostly in the central Andes of the Ayacucho Valley.[229] Another is made from manioc among the Napo Runa people, who live in the Ecuadorian Amazon.[230]

One Indigenous society that is synonymous with beer is the Rarámuri people, who live in the Sierra Madre Mountains of northern Mexico. They produce a beer called *tesquino* made from corn fermented with a local grass seed (*basiáhuari*) as well as other ingredients to increase the fermentation, sweetness, or medicinal qualities[231] (Table A.4). It is described as a "thick, milky, nutritious brew."[232] Seven days are required for brewers to dampen and sprout the grain, then grind, boil, and ferment the beer in a larger ceramic vessel.[233] The average brew is around 50 gallons, and because it will spoil rather quickly, it is best to drink the beer within 12 to 24 hours from fermentation. The Rarámuri compose their daily life and their religious and economic activities around the production and consumption of beer, which will be discussed in Chapter 6.

There is variation in production methods within Rarámuri when the availability of maize is limited before the harvest period with the use of either green corn or the hearts of the maguey plant.[234] Production with green corn occurs when the dried corn is about to run out and the new harvest is about ready. The green corn is mashed, cooked, and strained to make beer. Sometimes the corn stalks are crushed in water, boiled, and then allowed to ferment. Usually, the green corn and stalks are added to the dried corn to produce the beer. The maguey plant is baked in a pit oven and then crushed in water using a large oak limb in a natural rock basin. The juice is squeezed out of the pounded maguey and strained through thin agave leaves stretched between two sticks. The juice is allowed to ferment, and then the brewer adds it to the beer after the beer has been cooked.

Figure 2.6 Raramuri brewer preparing chicha.

Conclusion

Our knowledge of past beer production is becoming clearer, but there were many decisions ancient brewers made that we cannot detect even with the best archaeological scientific techniques. However, like a sailboat tacking back and forth with the wind,[235] using the archaeological record in combination with Indigenous brewing techniques of today, which inevitably are very different from those of our ancestors, we can have a more complete understanding of how beer has changed over time. European beer production is an excellent example that clearly documents how changes associated with gender, technology, economy, and culinary choices influence beer through the centuries. While change may not happen quickly, it does occur and can have a profound effect on the people who make and drink beer. Women brewers continue to make most of the Indigenous beer throughout the world; among industrial brewing operations, men dominate. The strong association of beer and women deities attests to the importance that women had regarding ancient beer production, whereby women possibly had a profound influence on the creation and change of beer production over many millennia.

The history of beer production throughout the world reveals dramatic diversity in the ingredients used over the years, and the innovative and varied techniques brewers have created to brew beers reflect their cultural identity. The advent of beer was once thought to be at the time when hunters and gatherers became farmers and domesticated grains and beer became even more popular with the development of state governments, such as in Mesopotamia, Egypt, China, and the Andean Highlands. Archaeological research focusing on beer has been a slow process, partly owing to researchers not taking beer seriously because they were imposing their own Western perspectives on the role of beer in their own society. However, we now know that beer predates grain domestication in the Near East, and there is growing evidence that this may hold true for other parts of the world. Archaeological and ethnographic research confirms that beer is one of the most important foods for many Indigenous societies around the world, both in the past and present, and may be one of the principal culinary symbols of ethnic identity that has existed at least during the past 10,000 years. The next chapter of this book addresses the role that beer has and continues to have on the health, economy, technology, and religion of societies throughout the world.

3

The Near East and East Asia

Funerary Stone Pits, Red-Crowned Crane Flutes, Ancient Hymns, and Bear-Hunting Rituals

For the last 13 millennia, people from the eastern shores of the Mediterranean, the hills of northeastern India, the central river valleys of China, and the northernmost island of Japan have been produc-ing and drinking wild and domesticated wheat, barley, rice, and millet beer.

The archaeological evidence from Asia suggests that we need to re-think our perceptions that the first beer production was associated with early agriculture, that bread was the motivating factor for our control of plant production, and that the first complex rituals and monument-building were associated with the advent of food domesti-cation and state-level societies.

This chapter will take the reader on a journey beginning 11,000 BCE in Israel when hunting and gathering communities produced beer from wild grains in stone pits adjacent to their burial grounds as part of rituals to honor their deceased.[1] Nearly four thousand years later in China, early farmers consumed beer while musicians shook turtle-shell rattles and breathed into flutes made from the wing bones of the red-crown crane, a bird later ritualized as representing immortality,[2] Five millennia ago in ancient Mesopotamia, scribes recorded ancient hymns that honored women beer deities and revealed the practice of kings offering beer to other kings in gift-giving ceremonies.[3] To trace the history of beer as an integral part of funerary rituals and as a central commodity, we witness the transformation of the human story from hunter-gatherer to farmer, to statesman whereby surplus foods, social stratification, and state-sponsored religion would change our world forever. Today, the development and spread of Islam and Hinduism have curtailed the production of beer in many parts of Asia. While

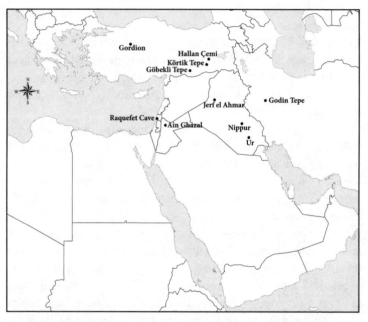

Figure 3.1 a. Map of the Near East—Map with sites of places discussed in the chapter.

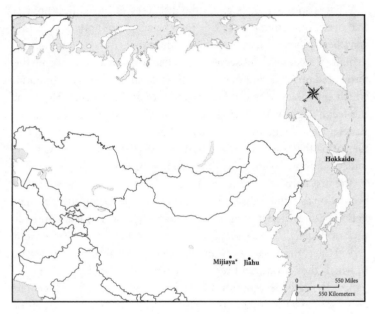

Figure 3.1 b. Map of East Asia—Map with sites of places discussed in the chapter.

there are many Indigenous societies in South and Southeast Asia who produce distilled alcohol from grain, they do not produce beer from malted grains. Nineteenth-century ethnographies document only one Indigenous society in Asia, the Ainu of Japan, who continued to use beer as a medium to touch the spirits, particularly during their annual bear-hunting rituals. However, the Ainu were beginning to be colonized by Japan in 1869, and their way of life has all but vanished since.[4] In this chapter, all of these examples, and many more, will be explored.

What is so remarkable about the discovery of the earliest brew is that beer production occurred before the advent of pottery. It has always been assumed that beer was produced when malted grain accidentally fermented.[5] Subsequently, someone decided to consume the fermented product, and they liked the taste and felt a slight euphoria, which encouraged the production of more of the product. However, the technology of producing beer in small stone mortars found in Raqefet Cave suggests that there was knowledge of brewing and fermentation that was not serendipitous but was the result of a structured understanding of plant processing that may have a deep history.

The first brew was created around 11,700 BCE before the advent of pottery or the domestication of grains in Israel during the Natufian Period.[6] The earliest brewers carved into the boulders and into the bedrock of Raqefet Cave in Israel to form pits in which they brewed and served the earliest beer. They lined the pits with tightly woven baskets in which they produced beer made from malted wild wheat and/or barley. The early beer pits are adjacent to a burial ground and seem to be associated with ritual funerary feasts that included the offerings of flowers, mountain gazelles, Mesopotamian fallow deer, wild boar, bezoar goat, aurochs, birds, lizards, and snakes placed within the human burials.[7]

The brewers at Raqefet Cave had a complex brewing technology in which they used each of the mortars (stone pits) for different functions. Researchers analyzed the starch grains, phytoliths, and fiber samples and conducted use-wear studies on the mortars to determine how the brewers used the mortars to make beer.[8] The starch and phytolith identification of the wild wheat and/or barley was done by using modern reference data.[9] Liu et al.[10] compared their analysis results with food- processing experiments and found that the grains from

Raqefet Cave were malted, fermented, and/or ground, and pounded, as well as attacked by enzymes.[11] Phytoliths, which are plant-based silica that have excellent preservation properties, were found in the mortar residue samples. Twisted and entangled fiber samples found in the mortars are most likely flax, which has a long history within the Levant dating to the Late Natufian Period (c. 11,700 BCE).[12] The flax possibly was used to make baskets to line the mortars for storing cereal malts.[13] Based on experimental studies, use wear on the mortar's interior indicates that the boulder mortars had fine linear striations created by the flax abrading the stone, and the rims of the boulder mortars exhibit a heavy polish with fine or furrow striations, all suggesting that the malt was being crushed or pounded in the mortars to make beer. Animal hairs also were found in the mortars, but it is not clear how they were utilized in the processing of beer. The mortars carved into the boulders most likely had stone slab lids, and these mortars were used for storing plant foods that included wheat and/or barley malts for brewing beer. The bedrock mortars had wider and deeper striations and were used to pound plant material with a wooden pestle for brewing beer.[14]

Brewers at Raqefet Cave were utilizing new technologies to process grains representing the vanguard of brewing. Concurrently, a Natufian site at Shubayqa I, Jordan, reveals the production of bread at 12,600 BCE, well before grain domestication.[15] Thus, this ground-breaking research from Raqefet Cave, in combination with the Shubayqa I site, suggests that both beer and bread may have been the leading factors in the domestication of grains in the Near East, since both sites predate the domestication of wheat and barley. After many years of research, archaeologists are beginning to answer the 1953 Braidwood and Sauer debate over whether beer or bread initiated food production. And as we will see, both beer and bread eventually propelled farmers to domesticate grains.

Long before the beer mortars were found in Israel, Hayden and co-authors[16] suggested that the numerous stone cups, bowls, and mortars at Natufian sites may have been used to process grains and consume beer. Elaborately decorated stone bowls during the Natufian Period or earlier (c. 13,000–9800 BCE) were usually made from basalt and were produced and transported some 60 to 100 kilometers (37–62

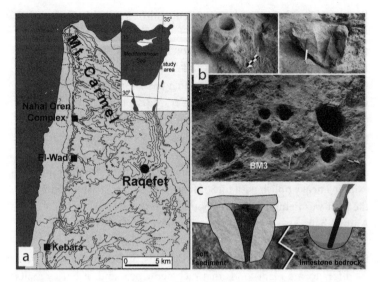

Figure 3.2 a. Location of Raqefet Cave; b. Raqefet Cave stone boulders and mortars; c. a reconstruction of how a brewer may have used the boulder mortar to store plants in a basket with a stone slab as a lid and a bedrock mortar used for brewing beer.

miles) from where the source material was quarried.[17] Evidence for these stone vessels appears at archaeological sites in Israel at Ain Mallaha,[18] in Syria at Abu Hureyra,[19] Jerf el Ahmar,[20] Tell Abr[21], and Tell Qaramel,[22], and in Turkey at Göbekli Tepe and Cayonu,[23] Nevalı Cori,[24] Körtik Tepe,[25] and Hallan Cemi.[26] Large, deep mortars found in Natufian sites would have provided brewers with the watertight containers they needed before the advent of pottery technology. Reed straws could have been used to draw beer from the mortars or used as part of the tightly woven baskets plated in the mortars.[27] Many of the stone cups and bowls have intricate incised designs and are thought to be symbols of prestige. These stone bowls show remarkable standardization with regard to their size, with rim diameters in a range of 9 to 12 cm and 8 to 9 cm in height.[28] The stone bowls with the decorative motifs would have taken such extensive time and labor for the artisans to produce that it seems highly likely they would have been used as serving vessels during important ritual feasts.[29]

Four Natufian sites—Körtik Tepe,[30] Göbekli Tepe,[31] Hallan Çemi,[32] and Jerf el Ahmar[33]—have evidence of these beautifully decorated stone cups/bowls for consuming beer. Of the four sites, the pre-pottery Natufian Körtik Tepe site in southeastern Turkey, dating from 10,050 to 9450 BCE, has the best examples of ritual-used decorated stone bowls.[34] Körtik Tepe was built and used by complex hunters and gatherers.[35] They built a nucleated village with houses made of stone walls with earthen floors, and they buried their ancestors under the house floors in a flexed position. The burials contained highly crafted incised stone vessels, thousands of stone beads, and stone axes, suggesting that there was social stratification at this time. Were these stone vessels used to serve beer during funerary feasts before the ancestors were buried?

To the west of Körtik Tepe in Turkey is an area known as the "Golden Triangle" in the Levant, which is situated in the middle and upper headwaters of the Euphrates and Tigris rivers and the foothills of the Taurus Mountains.[36] Here wild forms of einkorn and emmer wheat and barley grew, and wild herds of gazelle, boar, sheep, and red deer roamed the area.[37] Among these fertile fields of wild grain are Natufian preceramic sites such as Göbekli Tepe (9600–8200 BCE), which has remarkable ritual architecture and may have had early brewing associated with rituals. Four other similar sites are unexcavated, including Nevali Çorı, Sefer Tepe, Karahan Tepe, and Hamzan Tepe, which are now submerged under the Ataturk Dam.[38] Göbekli Tepe and communities across this region organized their settlements into distinct residential and workshop areas with open courtyards and evidence of widespread feasting.[39] Limestone vessels from Göbekli Tepe and Tell Abr 3 in Syria were large, with capacities of up to 160 liters (42 gallons).[40] Researchers believe beer was being produced in limestone basins using either wild barley or einkorn wheat, because they saw "grayish-black residues" on the bottom of the basins' interiors. In addition, it is probable that the samples they analyzed show calcium oxalate; however, results were not definitive.[41]

Göbekli Tepe is one of the most extraordinary sites, containing stone-lined circular structures with T-shaped pillars reaching up to 21 feet in height and weighing up to 50 metric tons, and other smaller stones erected in front of these giant pillars.[42] The amount of human labor needed to quarry, hand-carve, and then transport stone to the site is astonishing, and it is not inconceivable that beer was used as a food to energize and motivate the workers. Not only are the Göbekli Tepe colossal pillars notable for this time period, but they also have detailed etchings of foxes, lions, vultures, boars, deer, aurochs, wild ass, birds, snakes, spiders, and scorpions as well as hands, arms, decorated belts and a loincloth, and stylized humans.[43] Besides the zoomorphic and anthromorphic carvings on the pillars, there are sculptures of a boar and stone plates uncovered near one of the central pillars.[44] Ground-penetrating radar and geomagnetic surveys revealed that there are 16 other megalith rings at the site that have yet to be excavated.[45] It is not surprising that Göbekli Tepe is considered the "first human-built holy place" in the world.[46]

At the Natufian pre-ceramic site of Hallan Çemi in eastern Turkey, occupied from 10,200 to 7200 BCE, people collected wild barley, possibly to make beer for public feasts using stone bowls associated with burned-rock middens and sheep crania.[47] The elaborately decorated stone bowls and stone pestles were carved in a naturalized fashion, including goat heads, possible bovid horns, and mammalian animals lacking horns. Both the bowls and the carved pestles were made from chloritic stone, a black or dark green stone, and the pestles were particularly highly conserved, with the broken ends continuously carved to look like new.[48] It is argued that they both were used for preparation and consumption for ritualized public feasts.[49] While there has not been definitive evidence that beer was being brewed at Hallan Çemi, the large concentration of fire-cracked rock certainly could have been a by-product of extensive mashing using the highly decorative stone bowls found at the site.

The Jerf el Ahmar site in northern Syria, dating from 9500 to 9000 BCE, has evidence of wild barley in association with the basins and suggests that the grain was soaked to make malt and, eventually, beer.[50] Other wild charred cereals of barley, rye, and einkorn wheat were found on house floors and hearths.[51] In addition, almost 400 ground stones were uncovered, with 30 ground stones found in situ at 11 houses, and nine of the ground stones were found in groups of two, three, or four, suggesting that people were working together. Nine rooms were used only for grinding grains, and within the site there seems to have been semi-subterranean communal storage rooms for

(a)

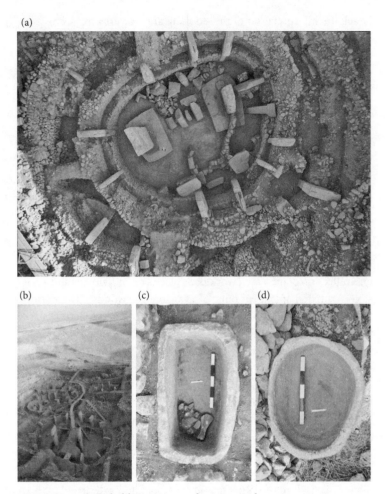

(b)   (c)   (d)

Figures 3.3a–d Göbekli Tepe site and stone vessels.

grain that were surrounded by houses.[52] One room in particular is interpreted as an area for processing and cooking foods, evidenced by the presence of three ground stones, hand stones, two large flat polished stone plates, one hearth, three limestone basins, and one small limestone bowl in association with wild barley, rye, and einkorn wheat. Specifically, with the discovery of brewing at the site of Raqefet Cave in

Israel, the interpretation of this cooking area for brewing seems consistent with beer production before the advent of domesticated grains or pottery in the Near East.

The discovery of beer production at Raqefet Cave, as well as the abundance of elaborately decorated and standardized stone bowls, strongly suggests that beer was produced before the invention of pottery. Before pottery, brewers could have relied on boiling the barley or wheat mash by wet-cooking using hides or barks. The brewers would have carefully elevated the container above the fire to bring a slow boil without destroying the organic container.[53] Another method would be using heated stones to dissipate the heat within the container and eventually causing a boil.[54] During the pre-pottery period in the Near East, people used clay balls to boil their food, as is found at the spectacular site of Catalhoyuk in southeastern Turkey.[55] The change from indirect to direct boiling did not alter what foods were boiled but may be a reflection of increased workloads placed on individuals and a possible reduction in the amount of available fuelwood.[56] The optimal time to boil mash, on average, ranges from 30 minutes to four hours.[57] Others suggest that because of the daytime heat in the Near East, boiling was not required to reach the optimal 65°C to 70°C for boiling mash.[58] However, even if the daytime temperature reached 120°F/ 49°C, it would have been warm enough for the mash to have a ß-amylase conversion (mash enzyme that degrades starch to create sugars for fermentation).[59] Once potters developed pottery technology, the output of beer production would have increased dramatically.

Later in the Near Eastern Levant, sites associated with early farming may have evidence of brewing. It has been suggested that the Ain Ghazal site in Jordan, dating to 7000 BCE, had floors designed for possible malting for the production of beer using domesticated barley.[60] The floors are colored with red plaster and highly burnished, with some of the floors consisting of plaster as thick as 14 cm.[61] The people living at Ain Ghazal were making and using materials, such as ground stones, pottery, ovens, or kilns, that could assist in the production of beer.[62] Ain Ghazal is famous for its small (~80–90 cm) plaster statues of individuals found in a pit of an abandoned house that was made with reed and twigs and covered with plaster.[63] A plastered skull also was uncovered from a pit located in a domestic area of the site dating to the pre-pottery period (7000 to 6700 BCE). Other plastered skulls have been uncovered from Jericho in Palestine and from the Nahal Hemar cave site in the Judean Desert.[64] What is the meaning of these exceptional symbolic creations, and could it have anything to do with beer production, feasting, and remembrance of ancestors who passed down valuable knowledge about living in the world at this time?

Early Natufian sites in the Levant strongly suggest the production of beer by hunters and gatherers, who processed and served wild barley and wheat beer at feasts in stone bowls, cups, and mortars. At these sites, the prevalence of limestone vessels and ground and hand stones for processing wild grains possibly indicates that this technology could have been used to produce beer, and this may have been associated with feasting, since the sites also have a high frequency of broken animal long bones, possibly to extract the marrow.[65] Many of the stone vessels are highly ornate with geometric patterns, and some bowls have images of animals.[66] Some researchers argue that individuals who hosted feasts with surplus foods may have encouraged the eventual domestication of plants and animals[67] or maybe a collective group used feasts to maintain ritual obligations to generate and uphold social relations.[68] Hayden[69] argues that food production was implemented in times of plenty, when control of labor and materials for mass seed-gathering, long-term storage technologies, and food/beer processing would provide opportunities for certain individuals to elevate their status. These feasts may have fueled a response to find new food sources such as the cultivation of cereals, which led to the eventual domestication of grain crops.[70] However, others view feasting as a response by individuals who wished to create and validate social networks through communal rituals.[71] Feasts built social storage between communities, creating alliances that could be called upon in times of food stress, thus mitigating periods of food scarcity. These feasts then continued to fuel subsistence activities to ensure that the ritual feasts endured through time as a mechanism to display, distribute, interact, and consume. Material culture associated with feasts may have had special properties and may have been procured from distant and symbolically powerful places.[72] Moreover, ritual exchange items may have been produced to enhance social networks, especially when there was environmental or social stress.[73] Whichever model one may wish to follow, the archaeological record strongly suggests that early brewing was being created long before the domestication of grains.

Red-Crowned Crane Flutes and Beer in Early China

Four thousand years after beer was first brewed in Southwest Asia, ritual leaders used beer as a food and possibly during funerary rituals in China 7000 BCE.[74] At the site of Jiahu, in north-central China along the Huai River Valley that leads to the Yellow River, farmers were cultivating rice and potters were making vessels to ferment beer.[75] *Hu*-type pottery vases found in the village of Jiahu were analyzed for their residues, and it was determined that beer was made from rice, grapes, hawthorn fruit, and honey.[76] *Hu* vessel types have been uncovered in both household and funerary contexts.[77]

What is also remarkable about Jiahu is that of the 349 burials found at the site, some contained flutes made from the bones of red-crowned cranes and engraved turtle shells. While it is not known if beer was being drunk at the time of burying these ritual leaders, we do know that beer was present at the site, and piecing together the archaeological evidence we can begin to indicate how beer and ritual objects may have intertwined within Jiahu society. The red-crowned cranes at Jiahu may be the earliest representation of these birds in Chinese culture; today they represent immortality.[78] A human mandible placed on top of the chest of one of the flute players suggests a gesture that may embody a linkage to the person's ancestors.[79] In another burial of a musician, people laid two flutes on either side of his left leg and placed the turtle shell with pebbles next to his right shoulder as well as a three-legged cooking jar, a vase, arrows and barbed harpoons, ground stones, awls, chisels, knives, and other items.[80] They cut the head from the body and placed the head toward the northwest, which was a common treatment for Jiahu burials at this time. All in all, 25 flutes made from the ulnae or the wing bones of the red-crowned crane (*Grus japonensis*) were recovered from Jiahu burial contexts. The red-crowned crane stands to a height of five feet and is the heaviest of all the crane birds on earth,

weighing in at a staggering 33 pounds.[81] Flying cranes in association
with cloud motifs have a long history in Chinese art, from funerary
tomb art from the early Tang Dynasty (618–906 CE) to silk garments
in the Liao Dynasty (907–1125 CE) to silk tapestry and beautiful hand-
crafted Koryo celadons during the Song Dynasty (960–1279 CE) to
wall paintings from the early Yuan Dynasty (1260–1368 CE).[82] In a text
written by an anonymous author possibly during the Song Dynasty on
the physiognomy of the crane:

> He is a yang bird yet roams in the yin world. It goes through various
> stages of transformations and takes one thousand and six hundred
> years to complete its final transformation. Its white, feathered body
> indicates that bird's pure and clean nature. The red crown on its head
> indicates that its calling reaches heaven. Its longevity is immeasur-
> able. While moving, they always stay on islets in a river, while resting
> they always gather in the woods. They are the senior leader of birds
> and vehicles for the immortals.[83]

Some of the world's first beer drinkers who lived and died at Jiahu
were buried with flutes from the red-crowned crane that may have
represented the longevity and immortality that would become a fixture
in later Chinese symbolism. While the time span between Jiahu and
the later dynasties is great, I do not believe we can discount the symbolic importance that the red-crowned crane flutes had to the people
of Jiahu.

Engraved turtle shells, perhaps representing the earliest writing
in China, were also present in 23 of the 349 burials, suggesting that
these burials were high-status individuals, whose deaths likely spurred
drinking festivities.[84] As with the symbolism of the red-crowned
crane that later became associated with immortality, turtle shells in
China became the medium from which writing was placed, usually
representing divination related to the dynastic royal houses.[85] Pairs of
shells with drilled holes may have been tied together to secure pebbles
found in the pair.[86] It is possible that the markings on the Jiahu tor-
toise shells are an early form of divination and that the pebbles may
have an early use of numerology or may have been used only to make
music.[87] Chinese burials dated later are richly adorned with many

items, including turtle shells, which have a long history of representing high social status.[88]

Jiahu was occupied during a time of transitioning to domesticated rice, and because they were processing beer as part of their daily calories and for ritual use, beer likely also assisted in the protection of the health of Jiahu residents over time.[89] The only serious health issue the Jiahu population faced was a lack of protein, which caused iron-deficiency anemia.[90] Over time, the burials indicate that the community became healthier and work became less stressful, lowering the risk of degenerative joint disease. Some Jiahu men suffered more from osteoarthritis than women, suggesting that men had more physically demanding workloads.[91] However, the Jiahu community members were surprisingly healthy in view of the fact that they were living in a settled community, which would usually cause serious health problems because of unsanitary conditions. The osteological analysis of their skeletons revealed an extremely low rate of periosteal disease, or bone infections, which are often caused by living in settled and cramped communities.[92] However, the earliest Jiahu settlement that was transitioning from hunting, fishing, and gathering to food production was small, and therefore the rate of potential disease due to unsanitary conditions was low. But more important, many of the vessels were most likely used to produce beer, which would kill most of the harmful bacteria in the untreated water.[93]

Four thousand years after Jiahu at around 3000 BCE in the Yellow River Valley during the Yangshao culture (5000–2900 BCE), there were three Yangshao sites where amphorae pottery vessels reveal through residue analysis the brewing of barley and millet beer (see Chapter 2).[94] The amphorae during the Yangshao Period became larger over time, measuring more than one meter (more than three feet) in height, suggesting communal drinking patterns and competitive feasts.[95] These feasts practiced throughout the world were meant to mobilize groups that would be economically, socially, and ritually tied to the leader to organize work labor groups, build social identities that would enforce economic surpluses and produce profits, build political power, and reinforce ancestral and deity connections to foster peaceful relations among social groups.[96] As Liu[97] outlines, this was a time of great cultural expansion in China, with the first development of a writing system,[98] the craft of making tools and ornaments from copper and bronze,[99] the construction of organized town walls, an increase in warfare,[100] and the development of social hierarchy.[101] All of these significant cultural advances led to the eventual development of state-level society.[102]

In China during the Zhou Dynasty around 1500 BCE, elites continued to hold ritual feasts, serving large amounts of millet beer. These feasts included a performance by the *shi,* or "personator of the dead," who contacted the ancestors and gained status by his performance. People ate a variety of meats, but the most important item on the menu was millet beer.[103] By the time the Han Dynasty (206 BCE–200 CE) was established, millet beer drinking and beer vessels were common offerings in elite burial contexts, and this continued into the historic period.[104] Contemporaneous with the dynastic powers of the Far East, Mesopotamia also would rise into a preeminent state-level society, where beer would be a central food for deities, leaders, and the common worker.

Hymns to Women Deities and Barley and Wheat Beer in Mesopotamia

Beer in Mesopotamia was a malted barley-based beer with an assortment of other ingredients from emmer wheat, date syrup, and "aromatics."[105] The earliest evidence of beer in Mesopotamia/Levant was discovered at the site of Godin Tepe from evidence of calcium oxalate found at the bottom of the interior of the ceramic pot.[106] Mesopotamia at this time, around 3400 BCE, was a society with incredible social and economic growth, with the earliest formal cuneiform writing system in the world, inscribed on clay tablets, and establishing the first code of law, the first irrigation system, monumental architecture in the form of temples and palaces, and the first bureaucracy. The production and drinking of beer was recorded on tablets found in Mesopotamia's early temples and palaces.

Mesopotamian writings indicate that drinking beer was perceived as a practice that humanized a person. The Epic of Gilgamesh, which dates ca. 2000 BCE and was part of King Ashurbanipal's (669–627 BCE) library from the city of Nineveh, states that what makes us human is drinking beer, as well as sex, food, and oil rubs.[107] In this story, Enkidu, who is strong and tall but a wild man with cattle hair, protects the shepherds but forgoes drinking beer until a sex worker convinces him to begin eating and drinking. Here is the passage from the Old Babylonian version:

> Enkidu knew nothing about eating bread for food,
> and of drinking beer he had not been taught.
> The sex worker spoke to Enkidu, saying:
>> "Eat the food, Enkidu, it is the way one lives.
>> Drink the beer as is the custom of the land."
> Enkidu ate the food until he was sated, he drank the
> beer—seven jugs!—and became expansive and sated with joy!
> He was elated and his face glowed.
> He splashed his shaggy body with water and rubbed
> himself with oil, and turned into a human.
> He put on some clothing and became like a warrior.
> He took up his weapon and chased lions so that the shepherds
> Could rest at night.[108]

This early writing is a testament to the importance of beer in shaping the world of Mesopotamians in which beer drinking was associated with becoming a respectable human.

Mesopotamian beer was associated with each segment of society, from the tavern where commoners convened to drink the latest brew to the deities and the cycle of the seasons tied to the production of beer. Mesopotamian beer taverns were another established social setting where beer was the alcohol drink of choice.[109] Taverns were usually placed near the town square and were considered to be places that could become wild and unruly.[110] However, they were places for everyone, including children and women, and served food and medicine, not just beer.[111] Priestesses owned taverns, and they were places where initiation rituals occurred based on texts from Sippar (1894–1595 BCE).[112]

Inanna was the goddess of war and sexuality, and a Sumerian poem describes Inanna's genitals as "sweet as beer."[113] Inanna was considered in Sumerian to be the most important goddess in Mesopotamia and closely associated with the city of Uruk.[114] Inanna falls in love with the shepherd Dumuzi, but in the end Dumuzi dies a mysterious death, and, depending on the story, he dies in a brewery.[115]. After Dumuzi's death, he descends into the netherworld, which is associated with the agricultural cycle. His time in the netherworld is connected to when the summer becomes so hot that all crops and other vegetation wither and die, but when the fields become green during the winter, Dumuzi returns to the world of the living. There are other cultural settings that intersect the behavior of the different deities and beer.

Beer in Mesopotamia was an important part of feasting whereby beer was a central food for the elite. Specific temples or ziggurats were dedicated to a deity and by 5500 years ago, each city had a patron god or goddess.[116] Other secondary temples were dedicated to other deities, such as in the city of Lagash, where according to texts there were 20 temples in the city. Deities could inhabit different temples or cities at the same time and take many different forms.[117] The deities would be represented as a statue, usually made from gold and wood, and although none survive today, texts indicate that the statues would be displayed during ritual feasts.[118] The deity statues would be clothed with fine garments and jewelry and fed every day with foods that would include beer, as well as bread, meat, fish, milk, cheese, butter, honey, and dates. The deities would travel around during special festivals so that larger populations could view their patron god or goddess.

Beer was a social drink for Mesopotamians based on seal impressions that show large banquets and feasts where men and women drink together from straws dipped into a large pottery vessel. It is rare to find a seal indicating someone drinking alone. Linguistically, it is interesting that the words kas-dé-a, representing a banquet in Sumerian text, are translated as "the pouring of beer."[119] In Uruk and other Ur III–Period cities, the banquets usually included large ceremonial feasts where royalty and other governmental officials celebrated specific military victories. One of the famous texts from the city of Larsa documents that

Figure 3.4 Mesopotamian Ziggurat in Ur, Iraq.

deities were given a variety of beers to enjoy, as well as food items such as bread and goat meat.

When Mesopotamian kings gave gifts to rulers of other territories, there was always a large ceremonial banquet where copious amounts of beer were served and lavish gifts such as silver rings, furniture, clothing, and precious metals were given out. These records do not go into detail about the amount of beer drunk, but they do clearly state that the king and his family "drank beer in the house of . . ."[120] The banquets are often discussed through the texts in religious terms, such as offerings of beer to Inanna in Iddin-Dagan's sacred marriage hymn of the city of Isin. Here is the translation of the text:

> They pour dark beer for her,
> They pour light beer for her,
> Dark beer, emmer beer,
> The fermentation vats bubble fervently.
> Mixing syrup with butter into a paste,
> Mixing butter into a convection,
> They make bread of syrup and dates for her.
> Beer at days' end, flour, flour in syrup,

(And then) syrup and "wine" at sunrise, they for her.
God and man go to her with food and drink.[121]

The Mesopotamian texts also signify drinking vessels and garments as symbols of royal patronage linking vassal groups with powerful kings most likely solidified with royal banquets offering ritual drinks and meals, complimentary toasts, gift exchange, and an understanding of the hierarchal relationship.[122]

Economically in Mesopotamia, beer was a commodity ration for workers, especially beginning in the Ur III Period (2100 BCE), when the great walled city of Uruk was at its height and state-level society was prospering in Mesopotamia.[123] At the height of urbanization, cities required a variety of commodities produced by workers who were working full-time and mass-producing goods. This growing force kept the cities flowing with goods and were fed partly by beer as compensation. Wealthy households used non-kin laborers to produce resources such as beer and textiles. Beer also was given to messengers and people with high status by the temples and palaces.[124] Beer may have been the sole drink for many people, since there were probably not many other suitable drinks that were produced on a large scale.[125] Mesopotamian texts indicate that substantial production and distribution of malt happened both in the household and within the organizational structure of the palaces and temples.

Foods made from barley, such as beer, bread, and flour, were the primary commodities dispensed by the temples, palaces, and private estates to compensate laborers once a month.[126] This tribute system of Mesopotamia was called *Oikos*. Cuneiform texts document an array of information, from receipts for purchases of raw materials and other products, to the selling and purchasing of land, to people's occupations.[127] The *Oikos* tribute system was made up of great households comprising temples, palaces, and estates who had a dependent labor force. The *Oikos* system was created to have "self-sufficient households whose members' labor provided for all of the household's economic needs."[128] Beer, made from dates, was given as rations to skilled workers.[129] Other foods, such as dates, honey, milk, cheese, meat, fish, and oil; were occasionally given out as well. The amount of the ration was dependent on age, gender, and the type of work that was performed in one of these great households.[130] Cuneiform texts describe the rationing of beer to a range of workers, from canal diggers, shearers, and farm workers to architects, messengers, and scribes.[131] During the Uruk Period, potters produced on a mass scale beveled rim bowls that were most likely used to distribute equal proportions to the laborers.[132] Besides laborers, others--such as royalty and gods-- received rations.[133] This may have been instituted to solidify a collective feeling among all community members, both rich and poor, even though rations for laborers were essential food (i.e., beer, bread, and oil) and for royalty supplemental foods (i.e., dates, milk, cheese, honey, and meat).

Different great houses during the Neo-Babylonian Period (625–539 BCE) were involved in brewing beer, and the employment of women to brew the beer was documented in the texts.[134] The great house of Egibi had a slave named Ishunnatu who received beer vessels, dates, and other brewing equipment to make beer, but she had to pay for the equipment from the sale of beer, with interest. She also had to pay the rent for her house, which she would turn into a beer tavern. This suggests that slaves during this time period were allowed to engage in substantial business activities, such as owning a beer tavern and brewing beer.[135]

The Code of Hammurabi, the mandated laws written for each citizen to follow, made sure that tavern owners did not overcharge their patrons, and if they did, women tavern keepers were to be drowned in the river.[136] During the Neo-Babylonian Period, the average quantity of beer ranged from one to three liters, with two to three liters being the most common.[137] This amount of beer fueled production of vast quantities of date beer by great houses such as the Murasu House in the city of Nippur.[138] Nippur is a city known for its temple of the highest deity, Enlil, and an educational campus known as "Tablet Hill," which contained more than 60,000 cuneiform tablets of Sumerian literature, including the story of the Great Flood and a map of the city of Nippur.[139] This amount of consumption would represent one of the primary foods for societies throughout Asia, thereby giving individuals the sustenance they needed on a daily basis. Beer continued through the ages to represent social hierarchy, as we see from the site of Gordion, where followers of the fabled King Midas possibly buried their leader.

The Rosette Wooden Serving Stand and Barley, Grape, Honey Beer in Turkey

Gordion, Turkey, located 90 km (56 miles) southwest of Ankara along the Sakarya River, was the capital of the Phrygian state (950 BCE– 150 CE), and during its height of complexity and population, people commemorated their dead by drinking vessels full of sweet beer.[140] A large earthen tumulus, where a chief or king was laid to rest, was constructed around 2700 years ago,. Accompanying the chief/king were the residues of beer inside a pottery jar made from grape wine, barley beer, and honey mead.[141] The pottery vessel was part of a large beer-drinking assemblage of 170 bronze vessels and 18 pottery vessels, as well as an array of exquisite wooden furniture. Three large bronze vessels contained pottery vessels that still preserved food residues from the mortuary feast celebrated 2700 years ago. The wooden serving stand found within the tomb was decorated with a rosette, a symbol associated with the Mesopotamian goddess Ishtar, who was the patron of beer, goats, and sex, which the Phrygians appropriated.[142]

Many believe this spectacular burial belonged to King Midas, who turned everything to gold by touching it. It is uncertain how the legend of Midas turning everything to gold came about, but it seems to symbolize the wealth of King Midas.[143] While not a single gold artifact was found in King Midas's tomb, the strong association of King Midas with gold most likely comes from the shroud that covered him, which contained an inorganic pigment called a goethite that had a golden appearance, suggesting the high quality of textile production among the Phrygians.[144] Based on the dendrochronology and carbon dates of 740 BCE, the tomb was most likely that of King Midas's father.[145] King Midas married the daughter of the King of Aeolian Kyme and was the first non-Greek to make a dedication to Delphi, where he placed a wood and ivory throne that Herodotus states he saw in the Corinthian treasury.[146] It is assumed that the Phrygians migrated into Anatolia/ ancient Turkey from southeastern Europe after the Hittite Empire collapsed around 1200 BCE. Eventually, the city would be conquered by nomadic societies from the northeast 400 years later.[147] However, the city of Gordion continued to exist under different conquerors, such as Alexander the Great in 333 BCE, but would eventually be abandoned.

The engineering feat of constructing a tomb that can withstand 53 meters of soil above it is remarkable.[148] Another remarkable aspect of the tomb was the writing of names on a wooden beam that was inserted into the tomb at the conclusion of the ceremony.[149] This is similar to today's tradition of signing a memorial book; however, to find this example with such antiquity is indeed rare.[150] The Phrygians built more than 100 burial mounds for a range of wealthy individuals, from elderly men to young boys, but not women[151]. Within one burial mound for a young boy, there were nine beer vessels, suggesting that beer was a substantial part of children's diet during the Phrygian Period. This should not be surprising, and we should not view this through our Western perspectives, since it is common to feed children low-alcohol beer in contemporary Indigenous societies.[152] The majority of beer pitchers are ceramic as well as bronze, with the ceramic vessel having elaborate geometric designs painted from the lower base to the spout and neck.[153] Other Gordion burial mounds included from seven to 15 spouted beer jugs, and the food that accompanied the beer included a spicy stew of barbecued goat or lamb, honey, wine, olive oil, and lintels spiced with anise or fennel.[154] While the tie between beer and the afterlife was an entrenched part of Gordion life, other Asian cultures have had a more complicated history with beer.

Uses of Millet, Wheat, and Barley Beers in India

Food traditions around the world, including beer, change as a result of societies, even individuals, interacting with one another. Sometimes this can have a dramatic effect on the culinary traditions within a region. In South Asia, this was the case with the introduction of millet beer. Vast trading networks on the Indian Ocean connected South Asia with the east coast of Africa, and this resulted in the spread of African millets to India around 2000 BCE.[155] The earliest state-level societies in the Indus Valley adopted a suite of African millets (i.e., sorghum [*bi-color*]; pearl millet [*Pennisetum glaucum*], and finger millet [*Eleusine coracana*]).[156] The Indus Valley farmers already grew and harvested Near Eastern crops of wheat and barley in the winter, and the adoption of African millets provided a crop they could harvest in the summer. With the adoption of these crops, social inequality seems to have emerged, with evidence of elite control and consumption of beer made from these grains.[157] The culinary beer tradition of India was probably similar to the African tradition of the pot and porridge cuisine, consisting of large open bowls and globular jars.[158] The globular jars were used for beer production in India, which is similar to the African pottery technology.[159]

Today, Hinduism teaches that one should abstain from alcohol.[160] However, the earliest Hindu scriptures written in ancient Sanskrit, the *Vedas* (~1500–1200 BCE), describe people enjoying drinking various forms of alcohol, including beer.[161] This was a time when the Indus Valley cities of Harrapan and Mohenjo-Daro were built based on cardinal directions tied to the celestial patterns of the sun.[162] These early architects organized the direction and placement of the city streets, houses, and public buildings. Large projects included building a city wall for reducing flooding and a mound on the west side that contained the bathing platforms for public washing and tied the rulers to their rituals. All of these structures were built with standard-size bricks. The cities were organized into administrative areas, granaries, and a burial area. People were buried with ceramic pots, stone amulets, and beads but typically were not buried with luxury items.[163] One of the most famous features of these early cities was a complex sewer system for the wealthier sections of the city, with covered drains that took sewer water away from the residential areas, which had interior toilets and showers.[164]

Text written on clay tablets was used to document the extensive trading that merchants were engaged in from the Macran Coast of southern India to the mountains of Afghanistan.[165] Unfortunately, unlike similar records in Mesopotamia, China, and Egypt, the text is still undecipherable.[166]

This state society was able to produce surplus crops, such as wheat, barley, and Asian and African millets, to make the beer because of the

Indus River flooding twice a year, allowing farmers to grow and harvest crops biannually.[167] Craft specialists, from brewers to potters, bead makers, stonemasons, weavers, and copper smelters, would have worked to provide the residents of the cities and people living in the hinterlands with goods that they needed on a daily basis.[168]

In ancient India, some groups were allowed to drink beer and others were strictly forbidden.[169] The first record of beer is from the ancient Vedic Sanskrit hymns of the Rigveda (c. 1200 BCE), which was used in rituals and was known by the name*sura*.[170] *Sura* was a strong beer made from either rice, millet, or barley as well as jaggary (crude palm molasses and blossoms of the mahuwa tree), and was drunk by all social groups, including the warrior class, the *Kshatriyas*, and the farming population.[171] *Sura* was drunk as a food and as a way to overcome daily suffering and hedonistic urges.[172] Drinking in ancient India was strictly enforced by religious and social constraints, with the high religious castes, such as the Brahmins, not allowed to drink the *sura* beer. However, other social caste groups were allowed to drink *sura*, and it was a common drink in public taverns.[173] *Sura* beer was also associated with ritual use, with *sura* vessels given to the king. However, *sura* and other drinks were considered evil as written in the Vedas and associated with anger, senselessness, and gambling.[174] Drinking *sura* is documented in the post-Vedic era with multiple references to drinking associated with soldiers and rulers.[175]

Sanskrit texts that discuss drinking during the post-Vedic era are difficult to date but may have been written around 300 BCE during the Mauryan Empire (321–185 BCE) and describe the state control of alcohol, including *sura*.[176] The Mauryan Empire exercised tight control over the production of beer and other alcoholic drinks, setting up government-controlled alehouses where it was the only place to legally drink.[177] Drinking was legal at this time, and the state profited financially from its taxation.[178] Special occasions, such as moving into a new house and weddings, were times that drinking was acceptable.

Early medical practitioners in ancient India (c. 300 BCE) were writing about the healthy aspects of drinking as long as it was in moderation.[179] Charaka, who was a medical practitioner, wrote in his medical text, *Caraka's Compendium*, that drinking alcohol had good and bad qualities.[180] He wrote a positive picture of drinking if it was drunk "by the right person, in the right manner, with the right preparations."[181] If one were to drink correctly, then the person needed to understand all of the factors involved before there would be positive effects. Charaka wrote about drinking moderately, stating that it is "pleasing, digestive, nourishing, and preserves intelligence."[182]

Another Sanskrit text, the *Kamasutra*, discussed the erotic nature of drinking and the pleasures it can bring.[183] The poet Asvaghosa described how the Buddha's father hoped that the Buddha would visit a pleasure garden where he would be distracted from his worries of growing old, disease, and eventual death.[184] Here women are drinking and are flirtatious with the Buddha, but because he is the Buddha, the hero, he does not succumb to these temptations. Another early text, the *Delight of the Mind* or *Manasollasa*, discusses the twelfth-century Kalyani Calukya, King Somesvara III, who experiences the game called the "game of drinking intoxicating drinks," where drinks are served in gold and silver vessels as well as glass that exhibits the colors of gemstones. The women at the drinking party become intoxicated and are lusting after the king, who is alone with the women. The text does not contain any evidence of immorality or unhealthiness but justifies the game as a way to demonstrate the king's exceptional ability to rule his kingdom.[185]

The acceptability of beer and other alcoholic drinks has had a complicated history in India with the adoption of Islam, which restricts alcoholic drinking.[186] Islam was the first major world religion to reject beer and other alcoholic drinks,[187] and this had a dramatic effect on beer production, distribution, and consumption, particularly in Asia. However, in Muslim courts in South Asia, drinking was permissible and sometimes quite common; autobiographies written by Mughal emperors Babur and Janahgir describe their propensity for alcohol by stating they were "great lovers of alcohol."[188] The Prophet Muhammad believed that people would drink fermented beverages excessively and that this would lead to immoral, sinful, and antisocial behavior.[189] The Qu'ran does not definitely prohibit alcohol consumption, but it does strongly suggest to abstain from drinking fermented beverages. However, there are two viewpoints regarding this matter. The most popular view is that no one should drink any alcohol, known as *khamr*, and the minority belief is that *khamr* pertains only to alcohol produced from grapes. Other fermented drinks, made from honey and dates, are allowed, but a person would still not be allowed to become intoxicated. There is also a split as to whether alcohol can be used for medical purposes, with the majority believing that it cannot, others believing that one may drink if their life is endangered with a medical condition.[190] In contrast to the complicated relationship to beer in South Asia, in the northern region of Japan, the Indigenous people of Ainu celebrated millet beer as a central part of their ritual life.

Bear-Hunting Rituals and Millet Beer among the Ainu of Japan

The Ainu, the official Indigenous people of Japan who live on the northern Japanese island of Hokkaido, offered millet beer during ritual feasts such as a bear ceremony and the Falling Tear feast to the bear and to the ancestors,[191] After 1868, Hokkaido officially became part of the Japanese territory, and dramatic changes began to affect the Ainu culture.[192] For example, they were forbidden to brew their sacred beer, were not allowed to fish salmon or to hunt deer or bear, and were pushed by the Japanese government to change from a hunting and gathering way of life, which was the heart of their religion, and forced to begin intensive farming.[193] The Ainu believed that the beer was sacred because it was analogous to a spirit and was brewed only for ritual ceremonies and the brewing was an inspiration as the brewers and consumers felt unity with their gods.[194]

Prior to Ainu annexation to Japan, ethnographers recorded their rituals associated with beer, such as marriage. Beer among the Ainu was a symbol of marriage unification.[195] A wedding gift given to the bride by the bridegroom was the ingredients to make sacred beer for the wedding feast. Prayers to the spirit *Kamui* were made, and the bride's father held a cup of sacred beer in his left hand and gently stroked the cup as he gave a poetic speech. He took the libation, and passed it to the bridegroom, who drank half. and the bride drank the remaining portion. As the bride drank, she made the women's ritual gesture by moving her right forefinger up her left arm to her lips and saying "*hap*," meaning thankfulness. Beer is the symbol that bonds the bride and bridegroom together, because if she does not drink the beer, then the marriage does not happen.

Beer was also associated with the Ainu bear ceremony, which was one of the Ainu's most important hunting rituals, first documented in 1710. The bear symbolized the reincarnation of the mountain god and sending the spirits (*ramat* and *kamui*) back to their world after they have visited the Ainu land.[196] One of the sacred objects used as an offering to the spirit *kamui* is an *inau*, which is a hand-crafted wooden stick the artisan whittles the end of the stick to produce a flowing bouquet of shavings. This winged *inau* was used in the bear ceremony to strain the sacred millet beer before the bear was sacrificed for the spirits. For the Ainu, millet beer is a symbol of status, since it was also only given as an offering to the village chief as part of his funeral but no other Ainu had beer brewed in honor of their life and death.[197] Unfortunately, the bear ceremony became a tourist attraction, and the Ainu were ashamed to continue on with the faux ritual because they believed it disrespected their ancestors and at the same time masked their discrimination, poverty, and forced assimilation by the Japanese state.[198] It was not until May 8, 1997, that the Japanese government officially acknowledged the Ainu as an ethnic minority.[199] This recognition by the Japanese government does not define the Ainu as an Indigenous people as delineated by the United Nations, and they remain culturally invisible to the larger Japanese society.

The feast of all souls or falling tears (*Shinurapa*) is a celebration, offering beer, to honor the ancestral *kamui* of the household by the living members of the household.[200] It is the only Ainu ritual in which women had direct communion with the ancestors. After food and drinks are consumed and offerings (i.e., two backward-shaven *inau*) are made to the *Kamui Fuchi*, a female ancestral spirit who resides beneath the hearth, a procession led by the women leaves the house with beer in a spouted vessel and a tray with an empty cup and four backward-shaven *inau* for the burial ground, where highly respected ancestors are buried. After salutations to the ancestors, the four *inau* are placed in the ground upright from north to south by the elder householder as he drips beer as a libation upon the four *inau*. He then prays to the ancestors, telling them that he has brought them beer, tobacco, cakes, and *lees* (beer residue), wishes the ancestors will give and watch over many healthy children, and then finishes the prayer by giving the four *inau* more beer as a libation. He feeds beer to his head-band and sword before drinking some himself, then passes the cup of beer to his wife. The wife feeds the four *inau* and pours some beer at her feet, drinks from the cup, then passes the cup to the next woman, the cup passing from woman to woman, who each drink from the cup before it returns to the householder, who returns it to the hearth. Prayers are made to the hearth spirit stating that they have followed her commands, then as more beer is drunk the husband sacrifices one of the backward-shaven *inau* to the fire. The women attending the ceremony continue to drink and pass the cup of beer to each other and pouring some on the ground to appease the ancestral spirits, and all the women make offerings of beer and food to the spirits in front of the sacred *inau*. Eventually, the food is divided among the celebratory members because the *kamui* had conveyed good power to the offerings and this allowed the consumption of the beer and food to bring health and longevity to the Ainu participants. The final act of the worship is each woman being given a backward-shaven *inau* by the officiating elder and each woman planting her *inau* into the ground and calling the name of an ancestress.[201]

After the solemnity of the worship, the atmosphere changes to one of a party setting, with the master of the ceremony sitting next to the sacred window with a vessel of beer, where he becomes the "controller of the drinking" (*sake-iush guru*, the person beside the liquor)[202] As the most respected elder, he moderates the behavior of the partygoers as he oversees the sacred beer, or *Tonoto Kamui*. Then the drinking, dancing, singing, and prayers begin, with participants usually drinking excessively, and with some dancing and singing around the beer container and the hearth.[203] The last dance is performed by the women as they mimic birds taking off in flight, possibly symbolizing spirits leaving the scene, since one elder stated that in the past the newly departed transformed into birds after death.[204] From the mortuary site of Raqefet Cave to the beer ceremony of Ainu, Asia has had a long and storied history of beer contributing to the efficacy of rituals. Beer was also a means for wealthy leaders to compensate the workers who built and embellished their world. Beer has played an important role in the Ainu's spiritual way of life, even as they have been portrayed in mainstream Japanese culture as primitive to attract tourists to visit their ancestral home on the island of Hokkaido.[205]

Conclusion

Asian societies have had a long and rich history of using beer as part of their rituals from the earliest brew, dating to 11,000 BCE at Raqefet Cave in Israel to the Ainu of northern Japan. During the transition from the Mesolithic to the Neolithic, when we often view societies as suffering from a range of health problems because of the lack of variety in their diets, the Jiahu community in China at 7000 BCE was in relatively good health. How people used beer through the ages and regions of Asia was as varied as the societies that inhabit this continent. While beer had a role in the development of food production and state-level societies that would impact the rest of the world, the deep history of beer in Asia now confirms that hunters and gatherers were incorporating beer into their lives and possibly used beer as a link between the living and the dead.

In Mesopotamia, beer was a symbol associated with deities, even to the point that the consumption of beer was tied to being considered a well-regarded human being. Temple personnel fed their patron deity beer and other foods as a daily offering. Beer was also drunk by the masses in beer taverns and was tied to gender roles. Mesopotamia had such an influence that later kingdoms appropriated deities associated with beer, such as the goddess Ishtar, who was symbolized on the wooden serving stand in the possible tomb of King Midas dating to 700 BCE.

After the onset of food production and when states begin to develop, beer becomes entrenched as a commodity used as a symbol of power to pay people at the household level as well as at palaces and temples. Once large urban areas developed, such as at Uruk, there was a need to feed workers who supplied the necessary crafts and labor to make the cities function, and beer was one of the leading foods supplied to the workers. This is especially true in the *Oikos* system, where great households, temples, and palaces had a large indentured labor force to feed. There was a demand for beer, with an average daily consumption of more than two-thirds of a gallon, so even slaves had the ability to manage a beer tavern as long as they were able to pay off the loan with interest.

In East Asia, societies such as at Jaihu were incorporating beer into their daily lives, as evidenced by their pottery vessels, and at the same time were burying their dead with intricate carved bone flutes from the red-crowned crane, a later symbol of Chinese immortality. Paralleling Mesopotamia, we later see the development of writing, feasts sponsored by the elite as they form their state systems. Other societies, such as the Ainu, historically have brewed millet beer as a symbol to pay respects to animal spirits such as bears and to their ancestors. Given the deep history the Ainu have had occupying the northern islands of Japan, these rituals have most likely been undertaken and changed over many millennia. However, historically we can appreciate how they use beer as a powerful libation connecting the living with their natural and spiritual worlds.

From its beginnings in the Indus Valley and the rise of Hinduism, India has had a complicated relationship with beer.[206] However, the beer *sura* was produced as early as 1200 BCE, and the state controlled the consumption of *sura* in alehouses.[207] In ancient India, the consumption of beer was considered a healthy practice as long as it was drunk in moderation, and it even had a role in aiding the mental health of the Buddha. However, beer represented caste, with Brahmins not allowed to drink because of their high social standing. Islam forbade beer in many Asian regions, resulting in the elimination of Indigenous beer in broad areas of South Asia, the Middle East, northern Africa, and southwestern Europe.[208]

Next, we discover the deep history of beer in Africa, a history older than even the pharaohs. Beer continues to pervade many African societies, providing economic capital, social prestige, daily nutrition, and a means to communicate with their ancestors.

4

Africa

Where Beer Feeds the Living and the Ancestors

Me: Calche, what do you have for lunch?
Calche: This is enough (he holds up a gourd of Gamo *farso* beer).

Africa, like the Near East, may have evidence for the oldest production of beer in the world, and it would most likely not be from domesticated grains, as the Braidwood and Sauer et al.[1] debate suggested. Archaeological evidence in Africa indicates that wild grains such as sorghum and finger millet were harvested well before they were domesticated.[2] These wild grains were so plentiful that early hunters and gatherers did not need to domesticate grains, such as what was discussed previously in the Near East. Africa has one of the oldest pottery traditions in the world, dating to 8000 BCE at the beginning of the Holocene,[3] possibly allowing the earliest African brewers to use their technological achievement to produce some of the world's earliest beer. What is most remarkable is the advent of wild grains with the greening of the Sahara and the Nile and the beginning of pottery production all, occurring at the same time. The Sahara was laden with rivers, marshes, and lakes from Sudan to Morocco, and there is evidence of beer processing from rock art in Libya, where an elder and a younger man are drinking from a straw out of a gourd. Use-alteration research, including residue analysis, is unraveling what these early vessels may have been used for. We know from organic residue research taken from the interior of the earliest ceramic vessels in the Sahara dating to 8000 BCE that they were used to process wild grasses.[4] The domestication of African grains does not occur until 2500 BCE, suggesting that beer

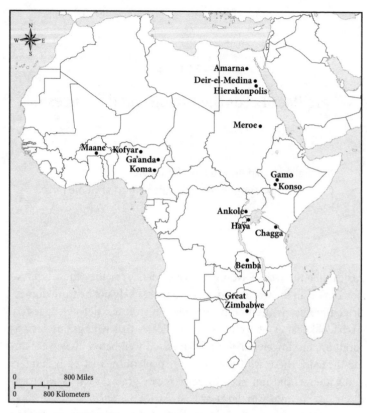

Figure 4.1 Map of Africa—sites and societies discussed in the text.

production was possibly in association with early pottery production and the harvesting of wild grasses.[5] Beer and porridges produced in ceramic vessels are African culinary traits compared to the bread and oven regime, which was so prevalent in the Near East.[6]

The culinary tradition of beer and porridges is evident in societies such as the Fur, who live in western Sudan, where beer is a staple food and where they begin feeding beer to their young at the age of eight months.[7] The quote at the beginning of the chapter describes my conversation with Calche, who was my landlord when I lived in the Gamo Highlands for two years. Calche enjoyed visiting new areas of Gamo when we would travel in our land cruiser to distant Gamo

Figure 4.2 Saharan rock art showing elder drinking with a younger man.

areas in search of potters and leather workers. Calche would always bring a gourd of beer for lunch, since this is the typical meal for Gamo farmers. The beer is thick, with a little more liquid than porridge and a low amount of alcohol, so the beer would satisfy him by filling him up but also giving him a slightly altered state. The ability to have beer available whenever Calche wished was a sign of status and power, since Calche was a wealthy farmer who had access to a ready supply of barley and had a wife who could do the intense amount of work required to transform the barley into a nutritious food.

Beer in Africa has a deep history, but the cultural significance of beer to each society is as distinct as each culture's history. How people have used beer over the millennia and centuries is a good barometer of how these societies have changed and continue to change, as well as of the uniqueness and diversity of Africa's societies. This chapter explores the deep history and contemporary use of Indigenous beer from the times of the pharaohs to communities today. Africans use beer in a myriad of ways, such as a source of income, to feed their ancestors for health and good fortune, to pay tribute to a leader, or to commemorate a wedding, house move, or funeral. While beer in Africa is produced and consumed in the majority of communities, each society reflects its own identity regarding its relationship with beer. For example, beer is a sign of status among the Gamo of Ethiopia.[8] It is used as a bridge to greet visitors and as a sign of friendship when someone visits a household among the Tiriki in Kenya,[9] to bond men together in Baganda, Uganda,[10] to settle disputes, discuss social issues, and tell stories among Tiriki and Iteso men in Kenya,[11] and as a symbol of renewal when a Nyakusa woman marries in East Africa.[12] These examples highlight the complex role that beer has played and continues to play in many contemporary African societies, structuring their social, economic, and ritual lives.

Pyramids, Pharaohs, Breweries, and Offerings

In ancient Egypt, wine was a beverage for the rich; beer was for the poor.[13] Beer was not only food for the workers, but the elites enjoyed beer as a staple and used it in sacraments to the dead. The Egyptian ontology or world view was cyclical and life started at birth, but a person was resurrected and reborn in parallel to the flooding cycles of the Nile and the patterned changes of the sun and the moon.[14] In the Book of the Dead, the pharaoh, on behalf of his followers, made offerings of beer to the gods and goddesses in the temples so that the deities would give life to the living.[15] The pharaoh gave Osiris an offering and states, "so that he might give bread, beer, meat, fowl (etc.) to the ka (spirit) of the deceased."[16] Feeding the deities beer meant that the deities would give to the dead, and therefore the reciprocal arrangement would allow the living to have access to the deities and the dead. The dead also could receive salvation by expelling their evil actions by approaching the Lake of Fire, which was a rectangular pit of fire with baboons on each corner and eight *uraeus* (Egyptian cobra) serpents. As the dead approached the lake, the evil dead were burned, while the blessed dead received nourishment from the lake. In the spell 126 dating to c. 1700 BCE, the four baboons speak about receiving nourishment, partly from beer.[17]

> Come, so that we may expel your evil and grip hold of your false-hood, so that the dread of you may be on earth, and dispel the evil

which was on earth. Enter into Rosetjau (the realm of Osiris in the Netherworld), pass by the secret portals of the West, and there shall be given to you a *shens*-cake, a jug of beer and a *persen*-loaf, and you shall go in and out at your desire, just like those favored spirits who are summoned daily into the horizon.[18]

Beer is first documented dating to 3100 at the site of Hierakonpolis BCE, the birthplace of Egyptian unification symbolized by the slate palette of King Narmer that memorialized his conquest of northern Egypt. This was a time when people from different cultural backgrounds living in the Sahara and southern Levant, who sought refuge from the mid-Holocene droughts, were migrating to the Nile Valley.[19] Over the next 1,000 years, people became more settled in their hamlets and began to become dependent upon the farming of emmer wheat and barley as well as fishing, hunting, and herding along the Nile. Division of labor increased between men and women to the extent that the Egyptian state was unique among early states, with its emphasis on the role of women in rituals associated with grain, beer, fertility, and death.[20] Since beer was such an important food for workers, it is not surprising that beer has been documented from the archaeological sites of Amarna and Deir el-Medina, where workers lived and drank the beer for household consumption. In addition, brewers living in Deir el-Medina produced beer so that they could place pots of it in the pharaohs' tombs after the workers constructed the tombs.[21]

In Hierakonpolis, Jeremy Geller[22] uncovered a brewery dating to 3100 BCE, where he found six pottery vats situated on a burned platform. In the interior of these earthen vats were thick (3 cm) black residues, mostly on the top and middle part of the vats' interior. When they conducted the chemical residue test, the chemist described the smell as "burnt brandy," revealing citric acid, succinic acid, and malic acid, all products associated with fermentation. The residues suggest a sugary rich by-product of mashing in association with uncarbonized and carbonized grains of emmer wheat that was accidentally spilled and burned in the fire to process the mash.[23] The aridity of the Sahara enabled this remarkable preservation of the wheat grains associated with the earliest known African brewery. The volume of beer from the six vats was about 100 gallons and a production of 300 gallons per week if the fermentation process took two days per brew. However, if the mash was transferred to other ceramic jars for fermenting, it is estimated that they could have produced 300 gallons of beer every day. Using the daily rations during the Middle Kingdom of 2.5 to 3 liters, it is estimated that the Hierakonpolis vat site could have fed 480 people per day.[24] The brewery was most likely run by the state to pay people in daily rations, since currency did not appear until the 26th Dynasty, some 2,500 years after the brewery at Hierakonpolis began.

Over time at Hierakonpolis, brewers first used pots with straw temper (i.e., the material that potters use to bind the clay together) and flat bases found at the Operation B brewery.[25] Potters may have used the straw to temper their pots from the byproduct of grain processing for beer production. Eventually, potters changed the design of beer jars, first to a conical shape and then finally to a pointed-base jar. The beer brewed at the Hierakonpolis supported ritual ceremonies in the nearby elite HK6 cemetery. One vessel (Pot 16) found at the Hierakonpolis brewery had a crescent-shaped potmark on its shoulder, applied by a potter when the clay was still wet. Similar pots with crescent-shaped potmarks have been found associated with graves in the elite cemetery.[26] Pot 16 had starch residues, indicating that it was used for beer fermentation, and it had evidence of calcium oxalate residues (beerstone) adhering to the interior wall of the vessel.[27] This comprehensive evidence strongly suggests that the beer jars that were fired in the Operation B brewery were filled with beer brewed from the brewery and then transported to a funerary feast and possibly other ritual ceremonies in the elite cemetery at HK6.[28] These beer jars would have been used mostly for transport, since they could contain up to six liters of beer, but would have been too large to be used for drinking vessels.[29] Rather, the black-topped beakers would have served well for drinking beer at one of the ritual ceremonies held at the HK6 cemetery. In one elite tomb (Tomb 16), 58 straw-tempered, crescent-shaped potmarked jars out of 115 pots were found in association with a number of burials of humans and a variety of animals, including an elephant.[30]

The world's oldest industrial-scale brewery, dating to c. 3000 BCE, has been located at Abydos, Egypt.[31] Egyptian brewers were producing

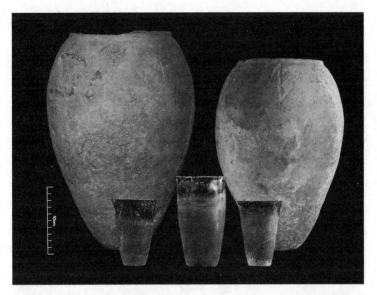

Figure 4.3 Hierakonpolis brewer's pots made with straw and black-topped beakers for drinking beer, found at the HK6 elite cemetery.

thousands of liters of beer to supply the funerary feasts of Egypt's first kings during the time when Upper and Lower Egypt were unifying politically and the Egyptian state was just beginning during King Narmer's time. Abydos had the first funerary temples that would lay the foundation for future funerary temples at Giza, Saqqara, and the Valley of the Kings and Queens. This funerary temple was located close to the brewery so that ritual offerings of ceramic beer jars could be placed within the king's funerary chamber. The brewery contained eight installations constructed parallel to each other, with each installation consisting of 40 large ceramic vats. Each brew could produce approximately 22,000 liters (5811 gallons). This amount would have provided a day's beer ration for more than 8800 people. The ceramic vats were for slow cooking the mash, and then the wort was strained and fermented. At the bottom of many of the vats, archaeologists found a "dark, glassy material" that consisted of the mash. The brewery was roofed, and there were stoke holes built into the side walls to fuel the fire to cook the mash.

Figure 4.4 Abydos Brewery showing the large vessels.

Figure 4.5 Mound of beer jars in front of the east corner gateway of the enclosure of King Peribsen.

Excavations led by Nadine Moeller and Gregory Marouard of the Oriental Institute at the University of Chicago at the Egyptian site of Tel Edfu have revealed an early Egyptian brewery.[32] Tel Edfu was occupied 2400 BCE during Egypt's Fifth Dynasty when the Great Pyramids were built. It was a place for workers who were engaged in mining of copper, gold, and jewels. The brewery was built using mud bricks and was situated near open courtyards and workshops. Clay seals found at the site show the King's royal prospectors engaged in their work. Copper smelting was being conducted at the site for the miners to have tools to mine their precious materials. Interactions and trade are documented from Red Sea shells and pottery from Nubia.

Beer as a form of payment for workers is first recorded during the pharaonic period, with the Egyptian pharaoh paying workers in monthly supplies consisting of vegetables, fish, firewood, vessels of beer, small cattle, and wine.[33] Every household produced their own beer, since the barley was paid to the workers to produce beer.[34] The workers also received beer from the pharaoh, and the quantity seems consistent in the texts as containing two jars. Payment was not always consistent, and we know that the first recorded workers' strike occurred during the twelfth century BCE or the 29th year of Rameses' III's reign at the village of Deir al-Medina.[35]

While the Egyptian culinary culture was based on bread and beer and agricultural crops of barley and wheat, the Nubians along the Middle Nile had a complex porridge and beer culinary tradition relying on sorghum and millet.[36] Historically and ethnographically, Sudan has a rich history of beer-making, with 20 to 30 percent of their grain production dedicated to beer production.[37] In the Kingdom of Kush at the administrative capital of Meroe, present-day Sudan, dating to around 100 CE, there are ceramic vessels from burials with the same erosion on the interior that I have found in contemporary Ethiopian beer pots, indicating that before these pots were put to rest with the deceased, they were used to produce and consume beer.[38] The remarkable Kush Kingdom had its own hieroglyphic writing that is undecipherable. It was first used to describe the accession of power through the matriarchy associated with the reign of Queen Shanakdakhete (c. 170–160 BCE).[39] Meroe negotiated a peace treaty with the Roman Empire after the Meroites attacked Aswan and captured the statue of Augustus, the head of which was discovered under the threshold of one of the Meroe palaces.[40] The kingdom had monumental architecture in the form of steep funerary pyramids, temples adorned with lion sculptures, and palaces. To feed the workers and the royal staff, barley, wheat, and sorghum were cultivated and processed to make beer. They mined for gold and semi-precious stones, as well as iron ore, so much so that it has been labeled the "Birmingham of ancient Africa".[41] While

Figure 4.6 Pyramids of the Meroitic kings and queens in the north cemetery. 300 BCE–400 CE.

some researchers dispel the view that this was one of the largest iron-production sites,[42] beer must have been an important food to replenish the workers' energy, whether they were tending their iron furnaces or agricultural fields along the Nile.

This reliance on beer is documented from the Nubian site of Meroe dating to 350–550 CE.[43] From 78 individuals interred at the Meroe site, it was found that all but four had tetracycline, an antibiotic, in their bones with no difference between age and sex of the individuals. Even two infants had tetracycline signatures in their bones, indicating that the mother was passing the antibiotic during nursing. The study indicated that the tetracycline in the bones was deposited not

Figure 4.7 Meroe funerary brewing pots.

as a one-time event of bacterial contamination of grains but occurred over at least 80 days of bone growth.[44] So where was the tetracycline coming from? Tetracycline comes from the bacterium streptomycetes, which is a mold-like substance occurring naturally in soils. George Armelagos and his team theorize that the streptomycetes spores were being captured in the brewers' warm, slightly cooked dough during exposure to the air to generate yeast. They conducted experiments and found that strepomycetes added to partially cooked bread and malted grains produced a significant amount of tetracycline.[45] Tetracycline is also found in ancient Egypt in the Dakhleh Oasis area dating to 400–500 BCE, which is contemporaneous with-the Meroe beer with tetracycline.[46] All of the individuals from Egypt and Nubia had no indication of bone lesions, suggesting that the drinking of beer with tetracycline had a therapeutic role in their overall health.[47] Besides providing health through a secondary source of antibiotics, ancient Egyptian beer had a wide range of medical properties: dressing wounds, treating gum disease by rinsing the beer in one's mouth or chewing and spitting it out, as an enema to treat anal disease, and as a vaginal douche.[48]

Other early African societies brewed beer, and we can tell this by the erosion on the interior of the pottery vessels caused by the beer's acidity, as I discovered and earlier discussed in the introduction. In Eritrea, people who occupied the pre-Askumite Ona sites that date to 400 BCE were the ancestors of what would become the earliest Christian state, known as Aksum. Here, to mark their burials people carved and erected single-stone monolith stelae that were taller than the Egyptian stelae.[49] The Ona culture is the precursor to Aksum and is located outside of the present-day Eritrean capital of Asmara.[50] The Ona sites, with beautifully crafted rock-lined houses, bull head figurines, and pottery, suggest beer brewing was a staple food for the people living there.

Great Zimbabwe: Where the Ancestors Are Silent

Great Zimbabwe was one of Africa's richest states, dating to 900 to 1450 CE, and is located in Zimbabwe near the Botswana and South African borders.[51] Great Zimbabwe and its earlier ceremonial sister city, Mapungubwe, were part of a large polity controlling vast economic wealth by trading cotton, gold, copper, glass beads, and tin and trading as far as the Far East, Eastern Africa, and Southern Asia. Gold, in particular, allowed the society to grow and prosper as surplus gold mining gave its leaders a new form of wealth that fostered the building of the colossal stone walls at Great Zimbabwe. The site reached its zenith during the fourteenth and fifteenth centuries, with a population expanding to 18,000 to 20,000. The city was composed of elite and commoner compounds and artisan workshops comprising more than 1,700 acres.[52]

Unfortunately, Great Zimbabwe over time had been used as a political and social pawn by the Portuguese, who could not fathom that this state-level city and regional polity was possible in Africa. German colonial explorers such as Karl Mauch attributed the site to King Solomon and Queen Sheba.[53] This fostered colonial control of the region if Europeans could appropriate the ancient site to European authority. Even though local clans lived on and around the

Figure 4.8 Aerial view of Great Zimbabwe.

site, colonial concessionaire companies began looting the site for gold and other precious metals, thereby systematically destroying much of Great Zimbabwe for future archaeological research.[54] Finally, after Zimbabwe became an independent nation, Great Zimbabwe's origin was rightly restored, and the site was established as UNESCO World Heritage site. However, the site and its interpretations remain politically charged between the local population and the post-colonial government.[55] Local clan leaders complain that the damage from colonial and post-colonial mismanagement has made the ancestors angry, causing a range of disasters, including brush fires, collapsing stone walls, and droughts.[56] Descendants believe that Great Zimbabwe is a sacred shrine where their ancestors still live.[57]

The spiritual world in Zimbabwe contains a connection between the older clans and offerings such as millet beer and tobacco snuff to Mwari, the divine provider of rain. Beer is often presented at sacred shrines in pottery vessels and gourds, such as in caves at Great Zimbabwe, where national biras (ancestral possession ceremonies) are held and goat and cattle are sacrificed for the ancestors to bring rain.[58] Thus, beer is tied to the rain-maker's ability to bring rain, life, and the sounds of the ancestors.

Symbols of Status and Feasting

Beer in African cultures is a critical component of the social, economic, and political structure of society.[59] Although people could process grains into bread, they spent a considerable amount of labor and time processing beer with which to establish alliances. Beer binds these different groups together by providing a means of establishing reciprocity in the form of labor and social and economic coalitions. Furthermore, the importance of beer in the form of symbolic respect to the living and to ancestors exemplifies its significant role in the well-being of societies.

The Gamo in southern Ethiopia, where I have worked for the past 20 years, have had a strict caste system, and the brewing and drinking of beer mirrors this social hierarchy[60]. There are three caste groups in the northern and central regions of Gamo and two caste groups in the southern region. Artisans, who are made up of potters (*chinasha*) and leather workers, iron workers, and ground stone makers (*degala*), are not allowed by the farmers (*mala*) to fully participate in political, economic, or religious ceremonies. Potters and leather workers in the northern and central sections of Gamo, respectively, will blow the horn to announce a meeting or ceremony, but they are kept from fully engaging in these communal ceremonies because they are considered impure. Beer embodies the ancestors because the ancestors bring fertility, and it is the ancestors who are fed beer at the beginning of each ceremony where beer is served. Thus, beer is a symbol of the fertility and wealth that are associated with the strict caste system both at feasts and in the household. While the drinking of beer unites the *mala* Gamo, it also segregates the artisans from engaging in Gamo ritual activities. Lineages belonging to the *mala* caste control the vast majority of farmland and produce the grains, such as barley, wheat, and corn, that are the essential ingredients for beer production. Large amounts of beer are prepared and served at political and religious ceremonies, such as initiation feasts for ritual sacrificers (*halakas*), weddings, and feasts celebrating saints' days, which are under the authority of the *mala* caste. There are communities within the Gamo region where artisans will elect their own *Halakas* and organize their own feasts, but never on the large scale of *mala* ceremonies. When the artisans do prepare these feasts, they must beg their *mala* patrons for grain to produce the beer.

Ritual sacrificers organize two beer feasts for this community of 300, which requires an astounding 800 kg (1,764 pounds) of grain, usually wheat or barley, for beer production.[61] To produce this amount of grain, the ritual sacrificer will rely on his clan, family, and friends to help supplement his grain production to produce this beer and other types of high-status foods. Beer feasts are symbols of the generosity the *mala* ritual sacrificers wish to bestow upon their community. The beer feasts are also ways for the community to reduce the accumulation of wealth by the ritual sacrificer and for him to turn this wealth into fertility for the community in the form of a beer feast.[62] Once the community decides that someone has achieved too much wealth, they will "catch" the individual, and he cannot refuse his new status as a ritual sacrificer.[63] Who drinks first is also a symbol of status, with the ritual sacrificer always pouring beer on the ground to feed the ancestors first and who have achieved the highest status by becoming ancestors. No one is allowed to drink before the ancestors are fed. Besides organizing the beer feasts, the ritual sacrificer sacrifices beer and animals to the ancestors in the sacred forests in which he prays for the health and fertility of all living things. Men serve as Gamo ritual sacrificers, who must be circumcised, married, wealthy, and morally respected.[64] Feasting is not the only context where Gamo people drink beer. The quote at the beginning of the chapter exemplifies that the Gamo consider beer a type of food and drink it in everyday affairs. Beer as a food is often consumed by the *mala* for their midday meal if they are traveling or farming, or they may purchase it at one of the weekly markets.

Archaeologists have paid a considerable amount of attention to documenting feasting in the archaeological record.[65] As with any type of analysis in archaeology, interpreting feasting is difficult. However, Wiessner[66] has outlined several helpful features of feasts. First, feasts are an aggregation of people, and this may be identifiable based on a large site or with the presence of a great diversity of artifacts. Second, feasts may be an indication of social inequality between hosts and guests. Third, feasts require abundance and the consumption of large quantities of beer, and this may be identifiable from households that are in charge of converting grains of millet, wheat, barley, or maize into beer. In rural parts of Africa today, oil drums and other metal vessels may be utilized to process beer, but in many places ceramic vessels continue to be an important tool to help produce African beer. There are many African ethnographic examples indicating that brewers use large ceramic jars for producing beer,[67] and this is also true of the Gamo, who use large jars that on average are about 34 liters/9 gallons in volume. However, the Gamo use large jars for other functions besides brewing beer. So how do archaeologists tell the difference between a vessel that was used for producing beer and one that was used for some other purpose? An analysis of the interior surface attrition and of context may help to provide the answer.

First, the spatial distribution of these vessels may offer information. It is common for the Gamo to congregate and conduct feasts within a ritualized meeting place called a *dabusha*. However, as most of the objects will be carried back to their respective households after a feast, it may be easier to identify such feasting by looking at the inventories of individual households. The higher Gamo caste households have the means to purchase the materials, such as large beer pots, grindstones, and grains that they either purchase or grow themselves, to produce beer for various feasts. Among the Gamo, beer processing occurs within kitchens and storage buildings, where large jars are placed along the outer edge of the interior of the building. Only the wealthiest households have the land and the capital to build these specialized structures. As discussed earlier in the chapter, the Gamo have an inequality between groups, with the wealthier, high-caste groups able to attend political and social feasts, while the poor artisan low-caste groups are excluded from attending. Gamo high-caste households have access to grains because of their monopoly on land ownership. This authority over grains enables the high-caste households to produce and consume more beer than lower-caste households. The feasting requirements for Gamo ritual sacrificers suggest that they should have the material culture indicative of the hosting of feasts. On the basis of the inventories recorded during my ethnoarchaeological research, I established that the ritual-sacrificer households do indeed have more vessels used for the fermentation of beer. In addition, ritual-sacrificer households have a higher frequency of gourds for drinking beer and ground stones for grinding grain than nonritual sacrificers' households.[68]

Because the character and size of ceramic inventories are archaeological signatures of beer production and consumption, beer production is also readily apparent in wear patterns within individual vessels. Among the Gamo, every vessel used for processing beer showed severe pitting or the complete erosion of the interior ceramic wall.[69] A possible explanation is that the beer-fermentation process produces lactic acid–forming bacteria that reduce the pH, resulting in a highly acidic substance.[70] The Gamo do not use utensils while producing their beer, so utensils are not the cause of this severe interior surface attrition. The pitting and/or surface attrition usually occurs from the rim to the base of the vessel. Although Gamo potters live throughout the region and use different clays, all the pots showed evidence of severe surface attrition, indicating that the fermentation process affects all clay types. Thus, this ethnoarchaeological research indicates that vessels used for processing beer have a higher rate of attrition and erosion of the interior wall than other vessels. Consequently, the presence of such surface erosion can be used as an important new indicator for the production of beer. By using my ethnoarchaeological research on Gamo beer vessels, other places around the world have documented the same type of erosion, including at the famous site of Paquimé/Casas Grandes in northern Mexico,[71] in burials from the site of Meroe in Sudan,[72] at Ona sites in Eritrea,[73] at the Neolithic site of Halai located on the North Euboean Gulf of Greece,[74] and in burials from the English Anglo-Saxon site of Cleatham,[75] and now I have found pottery from historic Gamo sites that are the original Gamo settlements with evidence of beer production and consumption.

Figure 4.9 Gamo beer pot with interior pitting.

The evidence of this case study also suggests that in some instances it might be easier to identify beer production in the archaeological record than beer consumption because beer consumption frequently uses organic vessels, which usually do not survive archaeologically. For example, the Gamo use gourds,[76] the Luo of Kenya[77] use a long vine-stem straw dipped in a large ceramic pot, the Zulu of South Africa use ceramic vessels (three different sizes) for drinking beer,[78] and the Iteso of Kenya and Uganda drink finger millet beer from ceramic pots using a siphon, but they use gourds or glasses when beer is made from either bananas or maize.[79]

Other African societies use beer feasts to gain status, such as the Koma of northern Cameroon, who brew sorghum beer. The beer provides one-third of the total energy consumed in a year and bonds age sets and establishes a hierarchy between each one.[80] The brewing of beer for ceremonial gatherings improves the person's status by allowing them to move up through the complex age grade system that will eventually allow them to become a high-ranking village policy-maker. To move up in the Koma age grade system, one has to produce copious amounts of sorghum to produce beer for the Nagë Napo, or cattle dance ritual. The ceremony can only be hosted if the community respects you and believes you have reached a level of status that warrants hosting this ceremony. The dance can be hosted up to seven times during a person's life span. At the last dance, the host will slaughter dwarf cattle and will be knowledgeable about all of the secret rituals and locations where sacred objects are placed. Once he has hosted the last dance, he has reached the highest status level and can drink beer from his own cup without sharing it with his fellow celebrants. He also will be buried with all the horns of the cattle he has slaughtered, signifying his prestige, and will return in the spirit world to protect the village. Hosting the cattle dance ceremony requires a significant amount of economic capital in the form of harvested sorghum to make beer. Besides the slaughtered cattle, the host must provide up to 75 pots of beer, which is the equivalent of more than 500 liters or at least 50 kilograms of sorghum flour to produce the beer. Although there is a considerable amount of drinking during the cattle dance ceremony and other gatherings, especially during harvest time, there is no hostility or violence, which would cause an individual to be shamed in front of the entire community.[81]

The Kofyar make, drink, talk, and think about beer.[82]

The Kofyar of northern Nigeria conduct all aspects of their daily life around beer and appreciate its taste and nutritious value as a food as much as its psychoactive effects.[83] Beer is shared from the same gourd among friends and lovers, and the giving of beer helps to cement alliances between peers. Harvest, hunting, and warrior celebrations, as well as dance and funeral rituals, all require large amounts of beer. Life-cycle changes also require beer: when a woman is breast-feeding, or when someone who has recently paid his taxes is given extra beer from the brewer who produced beer that day. When the Kofyar brew beer for sale, the producer is entitled to give a portion of the beer to the village head.[84] Diviners who have special knowledge of the future are initiated by a large beer feast, and as they travel throughout the hinterland are always given beer in recognition of their exceptional gift.

The spiritual role of beer among the Kofyar is as strong as in the living world. Great effort is spent appeasing the Kofyar ancestors that they are drinking beer rather than just gruel by blowing and pouring the beer on female and male ancestor graves and breaking beer jars when the construction of a grave cairn has finished.[85] The ancestors will then take the beer given to them by the living, provide it to the other ancestors, and thus receive high status in the spiritual world. Funerary rituals are conducted as a form of renewal to restore fertility and heal natural disasters by pouring beer on the ground. Spatially, the brew house within each Kofyar village is considered the most sacred area within each lineage's ancestral compound. Behavior near or within the brew house is sanctioned, since no one is allowed to quarrel or steal beer because this would mean certain punishment from the dead.

The Haya people who live in northwestern Tanzania along the great lakes pay respect to their fathers by offering them a gourd of banana beer, which must be done before others can be served.[86] The Haya pay their tribute in the form of beer, livestock, bark cloth, and iron products. With each brewing episode, the Haya must pay their king four or more gallons of banana beer. The Haya present the king with special gourds of beer that have wrappings of banana fiber covering the gourds and are tied with twigs and leaves from a plant that symbolizes purity and strength. As with the offering of beer to the deceased father, the Haya king offers a sacrifice to his ancestral altar for the welfare and fertility of his kingdom.[87] In addition, the Haya will sacrifice a gourd of beer at the ancestral altar to a father. The payment of tribute with beer indicates its economic and political importance, since beer is presented to kings as tribute among the Haya[88] and the Kofyar.[89]

In the Uganda kingdom of Ankole, when a man wants to seek permission from a prospective father-in-law to marry his daughter, they drink beer together out of a beer pot using a drinking straw at the potential son-in-law's house.[90] After the meeting, the drinking straw is placed on top of the house, signifying that the marriage agreement has been settled. After three months of seclusion by the new bride, who kept the house fire going, a feast is organized by the husband and then followed by a second feast from the bride's father. The drinking tube is taken down from the house roof and the husband, wife, and her father and mother drink beer together using this straw. The bride's mother then breaks the straw in pieces and stores the broken straw with other family treasures. Once the first child is born, the bride's mother makes a necklace from the straw and places the necklace around the child's neck. The straw and the beer represent unity between the two families, and the breakage of the straw and the re-creation of it as a necklace symbolizes the renewal and the different stages of life.

We have seen that death and beer are intertwined in some societies, and this is even more apparent with the Ga'anda peoples of northeastern Nigeria, where they respect death but view it as a dangerous time when the spirits should be controlled and celebrated.[91] The spirit of the recently deceased person is thought to live within a ceramic vessel, where it is kept alive for one year by cleaning the vessel constantly but also filling it with new batches of beer. The pot is kept in the deceased's sleeping room., which. After a year has passed, there is a second funeral, and the vessel is taken to the sacred forest, where it is held by the neck and smashed to release the spirit to the next world. This treatment of the deceased spirit promotes a positive relationship between the living and the ancestors.[92]

Colonizers, Religion, and Beer

Beer in Africa since the colonial times has been used as a medium to differentiate between people practicing different religions. Colonial missionaries, especially missions affiliated with Pentecostal religion, have instituted a no-drinking rule, and this has affected production and consumption of beer in specific communities. In the Gamo, the Pentecostal church has banned all drinking and smoking among its members, and this has affected the production of beer jars and water pipes for smoking tobacco in communities where people have adopted this new religion. In Gamo religion, beer is associated with feasting and for giving a sacrament to the ancestors to bring fertility to all living things. During the Derg regime in Ethiopia from 1974 to 1991, all Gamo religious practices and rituals were strictly forbidden; otherwise, individuals could be jailed or worse if caught. Many of the rituals were practiced at night when there was less chance of being seen by the Derg representative. When I lived in

Gamo, from 1996 to 1998, five years after the Derg regime was over-thrown, there was still a fear that lingered over the Gamo people when discussing anything related to their religion. Over the years, this fear has lessened, but the presence of the Pentecostal church has grown over the years. The Pentecostal church accepts artisans as part of their congregation, whereas the Orthodox Church has not, and this has resulted in a reduction in the production of beer jars among certain potter lineages.

Other African regions have witnessed religious changes that have caused a stigma about beer production and drinking.[93] In a Bhaca song from South Africa, there is a line that says, "Pagans are people of beer and beer alone."[94] The association of beer with ancestors and with drunkenness gave Christian missionaries a reason to push people to abstain from drinking and rise up "to a 'higher' Christian morality."[95] In Malawi, the Pentecostal movement has pushed for a rejection of past recognition of ancestors and an individual's connection to family and social and cultural responsibilities. This rejection of ancestors relates to the practice of giving libations of beer to the ancestors before ceremonies in return for benevolence, since beer and any other form of alcohol is thought to represent the devil and to promote and moral unrest.[96]

The ritual importance of beer production cannot be overstated, and the example from Burkino Faso highlights this. In Maane, the pro-duction of beer is tied directly to the ancestors, and therefore the re-ligious identity of an individual can strongly affect who may consume beer. For example, the Pentecostals do not drink beer and make sure they avoid any proximity to the rituals of beer brewing.[97] However, they are allowed to drink malt except for the malt produced during funerary rituals, because this would bring the Protestants too close to the ancestors, who would enact revenge killing on the Protestants for neglecting the ancestors. This avoidance of beer is contrasted with the Catholics in Maane, who brew and consume beer on a regular basis. They avoid the ancestors by not drinking from a gourd, which is used as an offering to the ancestors.[98] While beer may be under attack in specific cultural contexts, it remains the leading source of compensa-tion and helps to fund economic freedom for many different segments of the African population.

"Where There Is No Beer, There Is No Work"

Beer as a means of nutrition and compensation continues today in much of sub-Saharan Africa, and the subheading title, which comes from the Suri, who live in southwestern Ethiopia, typifies the importance of beer as a commodity to motivate the community to work on a variety of projects.[99] These work parties can be community-organized to build roads, bridges, and other infrastructure or could be brought together by a family needing help to plant or harvest their crops or move a house. In order to gather a work party, beer is essential, and, without beer, it is impossible to bring people together to cooperate on the task at hand. Work feasts are a type of commensal politics,[100] where people are organized to work on a specific project and then are served beer for their labor. Beer in work parties provides a social alliance between the work-party sponsor and the people involved in the work party. Beer changes the context of the group by becoming the "social focus" for all of the work-party members but also for people associated with the party members.[101]

The quality and quantity of beer as a motivational force to gather people for a work party is seen among the Konso of southern Ethiopia.[102] If a person produces a high-quality beer in sufficient quantities, then they can expect a range of 20 to 50 people to help them in the fields. There are a number of different types of work parties, which involve fixed or volunteer groups, costing the sponsor approximately 50 to 150 Ethiopian birr (US $7 to $21), depending on the size of the work party. Some in Konso can afford to organize large work parties, but this requires vast quantities of beer.[103] One Konso woman needed labor to reconstruct a wall that had been damaged by floods, and she was able to bring a group of more than 100 people from two Konso communities because she had the resources to produce enough beer to feed everyone throughout the day.

Feasts also accompany work parties, as in southern Africa, where chiefs and wealthier commoners organize large work parties and then provide an abundance of beer.[104] The Pondo of South Africa rate beer feasts higher than meat feasts because beer makes the work seem more like a party.[105] Among the Kofyar of Nigeria, beer is the primary means of repaying voluntary labor to hoe and harvest agricultural fields and to build corrals and houses.[106] The Kofyar treat beer as money used for reciprocity as part of a bride service and voluntary labor in the form of large work parties for preparing the fields, harvesting, and building corrals and houses, for example.[107] Having beer as a commodity helps beginning farmers because they can pay for volunteer labor as long as they can brew enough beer as payment. Individuals can also brew beer and sell the beer at a market and then purchase clothes, animals, and other commodities. If the brewer has to purchase grain to produce beer, it is almost certain that the brewer will make a profit from the selling of beer.

The Bemba, who live in Zambia, grow finger millet and process millet into beer to organize communal work parties for men to take part in the shifting cultivation of slash and burn (citemene) agricultural production or for women for collecting and piling up wood.[108] Bemba women process the finger millet into beer not only for labor recruitment but also to sell for other purposes.[109] Colonial governments espoused the position that beer brewing was wasteful, and women have in the past stated that beer brewing was on the increase. Beer brewing is one of the most profitable activities for women, and when other types of food are scarce, women state that their husbands subsist solely on beer. Joining a work party is one way for someone to access beer without diminishing their own food sources through the beer they will be provided after the job is completed.

Beer Brings Economic Freedom

Indigenous beer produced at the household level throughout Africa during the precolonial period was a seasonal food celebrating the time of harvest and utilized until the grains were spent and the famine season began.[110] In good years, there was a grain surplus when the beer production could continue throughout the year. Home brewing in many parts of Africa is one of the largest cottage industries, providing 7 to 20 percent of the employment in countries such as Botswana, Burkina Faso, Kenya, Uganda, and Zambia.[111] The introduction of high-yielding corn has added to or completely substituted for Indigenous crops, such as millet or sorghum, with regard to how beer is produced in rural parts of eastern and southern Africa.[112] Women in rural areas used corn to speed up the brewing process and allowed them to add extra household income.[113] This income allowed women to earn money for clothing and other household items but, importantly, gave women the means to invest in their children's education. This transfer of beer profits toward education occurred after the 1950s in the Nyakyusa society living in southwestern Tanzania, where women would spend the little extra income they made from brewing Indigenous beer to pay for their children to attend the government school in Tukuyu and mission schools, and when education became free in the 1960s, the earnings could pay for books and school clothes.[114] One woman interviewed by Justin Willis[115] explains, "I started because my mother had died and I did not have anybody to help me. I knew life would be hard with the children, so I started making beer." This cultural change of women making and selling beer caused stress in the household, because in the Nyakyusa society the husband traditionally controlled millet and commissioned his wife to make beer in order for him to build social storage in the form of obligations.[116] The brewing of beer by women for their profit was seen as undermining the control of men. As one man stated, "What brings disrespect is beer."[117] Women brewers moved to clubs where they felt safe to sell their brews, were not expected to sell on credit, and could control the profit. Nyakyusa women continue to produce beer to increase their household income, but it is hard work and profits are small and there is always the risk of a bad brew or not selling their investment. However, it has changed their lives by giving them more economic freedom to help themselves and their children.[118]

While beer production continues to help foster economic growth for women and their children's education, there has been a concern about the environmental problems caused by fuelwood consumption for beer production.[119] The amount of fuelwood and grain production for community feasts can have a considerable impact on the surrounding environment as well as taxing farmers and brewers to produce enough beer. For example, in the West African country of Burkina Faso, the Manga people organized five beer feasts in one week, which required 1,400 kg of fuelwood for brewing for just one feast and the processing of 1900 kg of red sorghum to feed the community during the week.[120] In South Africa, it is estimated that to brew one liter of beer, a brewer needs one to two kilograms of fuelwood, and for an average sub-Saharan rural community, the amount of fuelwood used to brew beer falls between 5 and 30 percent of annual consumption of fuelwood.[121] There is a myriad of recipes throughout Africa for brewing beer, with different amounts of fuelwood used, such that the banana beer *mbege*, which is brewed in Rwanda and Tanzania, requires five to six hours of brewing but uses much less fuelwood than sorghum beer.[122]

For rural African women, brewing beer adds substantially to the household income. For example, in Burkina Faso beer brewing is the third-highest household income source after farm sales and wage labor,[123] and in parts of Tanzania can be the second-highest source of household income,[124] with up to 75 percent of households brewing beer. Other countries have a high rate of rural brewing, such as in Zambia, with some households indicating that brewing beer is their first or second main source of generating income.[125] In Botswana, 20 to 50 percent of the households brew beer from one to seven times a week and beer brewing is one of the top income-generating activities for the household.[126]

Dietler's[127] interpretation of Gutmann's[128] account of the Chagga of Tanzania indicates that the chiefs gave generous supplies of beer, which fulfilled their redistributive obligations and supported the warriors who would fight on their behalf. The Chagga chiefs would collect this beer through a tax that the people were happy to pay so that they might socialize with the chief. The Chagga chief would also organize work parties to plant millet from which beer would be eventually brewed. A sign of a good Baganda chief is redistribution to the people in the form of "beer, meat, and politeness."[129] As with the Chagga, the Baganda chiefs obtain their beer through a tribute system.[130] Thus, the payment of beer forms a social, economic, and political reciprocal bond between commoners and leaders. While beer provides a critical economic commodity, it also has a rich history of providing essential nutrients and calories to many societies living in the past and present.

From Antibiotics to Vitamins

Numerous other African state societies incorporated beer as part of their ritual, economic, and daily lives, each with their own cultural identity. With all the dramatic cultural changes, including beer production and consumption, that have occurred over the centuries throughout the continent, African Indigenous societies continue to subsist on beer to foster their ritual, economic, and daily lives. African farmers grow one-eighth to one-third of their grains for processing beer, revealing the prominence of beer to many African societies.[131] The Zulu king Cetshwayo in 1881 listed to the Native Laws and Customs Commission that the Zulus consider beer a food.[132] Historically, it was reported that Malawians used 12.5 percent of their grains to produce beer, whereas in the Longone Plains of Chad it rose to 33 percent[133]. Other societies a century later continue to subsist largely on beer, with Botswanan households processing 15 to 20 percent of their sorghum into beer.[134] In Burkina Faso, it was reported that 50 percent of household grains were used to process beer, and in Kenya and South Africa it was 6 and 9 percent, respectively[135]. Even when the alcohol content is low, people continue to enjoy the sour taste and the thickness of the beer.[136] Botswana beer consumers state that they like the Indigenous beer because "It fills my stomach" or "It keeps me strong".[137] Calorie consumption from beer ranges from Zulu men in South Africa and the Tutsi in Rwanda and Burundi living off of sorghum beer, to Malawian men consuming 35 percent of their daily calories from beer, to Pedi men who rely on beer for 13 percent of their annual calories.[138] Sorghum beer provides one-third of the total energy consumed in a year among the Koma of northern Cameroon.[139] In some African countries, the per capita consumption of beer rivals that of known drinking regions, such as Germany; for example, the amount of per capita beer consumption is 347 liters per year in South Africa, 285 liters per year in Botswana, and 230 to 470 liters of beer per year in Burkina Faso.[140] To put this in perspective, Germans drink per capita 300 liters of beer per year, and in the United States adults drink on average 160 liters of beer per year. The consumption of beer provides much- needed nutrients, including calories, protein, B vitamins (including thiamine, folic acid, riboflavin, and nicotinic acid) and essential amino acids, especially lysine, that are often not found in grain-based African diets.[141]

Even during the most difficult times in people lives, such as during the twenty-first century when two million people worldwide have been displaced because of government-induced development, beer production and drinking continues as a way to ritually bond people together as they feed themselves with beer and to stimulate their local economy to better their lives[142]. Recently, there have been severe disruptions in southern Ethiopia because of the damming of the Omo River, which has forced families to move from their ancestral lands. Whereas their households were close enough to engage in daily activities such as sharing coffee, with the present community settlement the households are too far apart for people to easily engage in daily gatherings. The newly displaced complain that the land is treeless and there is no shade, so farmers who are part of the working party and are being fed in beer are now forced to seek shade under a small cloth tied to the corn stalks.[143] The complete disruption this plays on individuals, families, and communities cannot be overstated, but the fact that beer remains a staple as a daily ritual that pervades their social and economic lives is a testament to the importance that beer plays in their lives.

In many African societies today, whether times are good or bad, beer is a highly desirable food that socially binds people together and serves to reinforce social hospitality and communality during ceremonial and everyday activities.[144] Beer represents a common cultural marker of wealth and status; it is a commodity of reciprocity, hospitality, and communality; it may represent a payment of tribute to chiefs and kings and is an essential food in the redistribution of social and economic wealth. The processing and consumption of beer pervades many cultural acts, and, because of its social, economic, and political value, it is of great significance, both as a dietary staple and as a high-status food.[145]

Conclusions

When the Sahara was saturated with rivers, lakes, and marshes around 10,000 years ago,[146] beer may have been a central food for hunters and gatherers. Future research will determine whether this is true, but we do know that beer began to play a special role along the Nile River beginning at least 5,500 years ago. The dead are fed beer by the living, and this has a long tradition in Africa beginning with the pharaohs offering gods and goddesses beer so that they would bring life to the living. When I conducted my ethnoarchaeological research in southwestern Ethiopia, I realized the significance of beer as a bridge between the living and the dead. The Gamo ritual sacrificers feed the ancestors in exchange for health for all living things. Beer can also act as a gift of life, with the recently deceased being given beer as they approach the Lake of Fire in ancient Egypt as a symbol of eternal life. Beer was more essential for the general public living in ancient Egypt. The pharaohs had the workers place large pots of beer in their tombs so they could enjoy beer in the afterlife. In essence, beer is a symbol of power in many African societies.

In many different African societies, beer binds people together but at the same can be used as a symbol of power and status and thereby precludes others from enjoying beer as a daily or ritual food. For those who are allowed to drink beer, it is a staple and can provide a significant amount of daily calories. Today, beer is a means for women to gain economic power and control of their lives by giving them access to capital. Beer gives these women the opportunities for economic freedom that they can then transform into the education of their children. Therefore, beer in some societies is changing the lives of the next generation.

Great Zimbabwe offers the ideal setting, where beer not only links the descendants to their ancestors but serves as a way to right the wrongs of the colonial past. The ancestral possession ceremonies, or biras, are a way to present beer to the ancestors so that they will reciprocate by bringing fertility in the form of rain.[147] Other regions in Africa have witnessed changes in how beer is perceived and utilized with the onset of colonial governments and new forms of religion. Changing governmental philosophies can completely disrupt religions, as was illustrated when the Derg regime came to power in Ethiopia and banned all Indigenous religions. Competing religions also can change how people use beer in their rituals as Pentecostal missions take root in many parts of sub-Saharan Africa. Beer is a symbol of power when it is associated with the ancestors who bring life to the living.

Workers, today and in the ancient past, use beer as a food to fuel them in their hard labors. Ancient breweries were constructed to feed Old Kingdom miners as well as iron smelters living in Meroe of Sudan. People today use beer as a motivator to participate in work parties building roads and bridges, moving houses, preparing farmland, and planting and harvesting crops. These work feasts help to build relationships and bond people living in the community. Beer also continues to be part of people's lives when they are dispossessed of their land, because it provides important calories and nutrients as well as helping people to cope together in their daily struggles. The pervasiveness of beer in past and present African societies in the construction and maintenance of social, economic, and political relationships suggests that Indigenous beer will continue to feed both the living and the dead.

The next chapter takes us to Europe, where beer became an offering to large-scale burials as well as communal festivities, and was a major economic driver that eventually developed into beers that would span the world.

5

Europe

Ancient Henge Rituals, Beer Beakers, Celtic Funerary Urns, Vikings, and Witchcraft

Though Europe cannot claim the earliest production of beer, today it may have the greatest of beer styles, from ales to witbiers. Europe's earliest evidence of beer comes from a series of sites, mostly from Scotland (2900–1400 BCE) and Spain (c. 2400–2500 BCE).[1] The primary grains for brewing beer, eikkorn wheat and barley, were introduced from the Near East and Anatolia.[2] These crops first appeared in southern Europe c. 7000–4650 BCE,[3] Greece,[4] and Italy.[5] Central Europe farmers began utilizing these grains, reaching central Europe ca. 4450–2450 BCE[6] and Britain ca. 3600 BCE[7] before finally making their way into Neolithic Scandinavia ca. 3000 BCE.[8]

The movement of people from southwest Asia brought technological knowledge in the form of beer production and the bell beaker vessel that would typify the earliest beer culture in Europe.[9] During and after the Bell Beaker Period (2750–1800 BCE), a number of subsequent interactions across Europe occurred between northern and southern Europe.[10] These interactions would influence the way people would produce and drink their unique beers.

The variety of ingredients European brewers used over the past five millennia documents the ingenuity and sustainability of beer production there. Some have even proposed that European brewers used psychoactive ingredients that may have induced hallucinatory properties.[11] Some ingredients, such as rye, could become contaminated and have grave results, and was used by brewers to bring about hallucinogenic visions.[12] Eventually, hops would become an essential ingredient for taste but, more important, as a preservative for distribution of beers to other European and international regions.[13] European brewers would develop a range of technologies that would allow them to brew

Figure 5.1 Map of Europe—Map of sites discussed in the chapter.

both top- and bottom-fermenting beers, and their unique tastes would be similar to those of the variety of beers we enjoy today.

Farmers, Henges, and Beer in the United Kingdom

When one thinks of European archaeology, Stonehenge, the iconic World Heritage site with its magnificent megalithic triathlons located in Salisbury Plains, instantly comes to mind.[14] Built by early farmers about 2500 BCE, Stonehenge is not the only European megalithic structure; it is part of a larger complex of contemporaneous sites across

Europe that also contain monumental earthenworks and timber circles.[15]

> archaeologists have thought that the Beaker people also introduced al-
> cohol to Britain, although it was probably already widely in use in the
> form of beer and possibly cider.[16]

How was beer associated with Stonehenge and other megalithic monuments around Britain and Europe? One has to wonder if beer was used as a motivating force to organize the labor force to prepare and transport materials such as earth, timbers, and stone as well as to construct these monumental achievements onto the landscape. Beer also may have been a central agent in the rituals associated with the construction sequence of these monumental features, as well as during the feasting rituals aligned with the cosmological movements of the summer and winter solstices.

Direct evidence of beer remains elusive among the Neolithic large henges, such as the Stonehenge–Durrington Wall complex.[17] Although there is no direct evidence of beer production or consumption at Stonehenge, scholars suspect it was brought to the area by people traveling to attend feasts.[18] An avenue from Stonehenge aligned to the midsummer sunset also connects Stonehenge to the habitation and feasting site of Durrington Walls.[19] The site has northern and southern timber circles about 470 to 480 meters across placed inside of a large circular earthenwork.[20] Feasting is indicated at the site by a large number of pig bones, Grooved Ware ceramic vessels, and other culinary artifacts deposited rather quickly.[21] Unfortunately, there is no evidence of charred wheat or barley grains, ground stones, or beer residues in the Grooved Ware vessels.[22] Still, Craig argues that it is likely that beer was brought by people traveling to a feasting site, since we know that beer did exist at this time (c. 3800 BCE) in other parts of Europe (see later discussion). The pig and cattle bone elements, with no evidence of neonatal bones of either cattle or pigs, suggests that the animals were brought to the site from other regions and were not raised at Durrington Walls.[23] Strontium isotope analysis of the cattle remains suggests that the cattle were raised in the west of Britain, such as in Cornwall and Wales, as well as in northern Britain.[24] If animals were herded from distant places to be slaughtered at the site, then beer may have also been imported to Durrington Walls. One of the possible foods for the pigs raised for slaughter at Durrington Walls may have been spent grain from beer production as fodder.[25] However, some counter this, since the pigs would still be suckling milk and would not need spent grain fodder for food in their first year of life.[26] Until we find archaeological evidence, we will have to continue to wonder if beer was part of their ritual traditions.

Other sites that may have had elaborate feasting associated with megalithic monuments were at the Barnhouse village in Scotland, dating to 2600 BCE, which may have evidence of a malting room with barley husks in Structure 2.[27] Structure 2 at Barnhouse settlement was

Figure 5.2 Durrington Walls showing the wooden henges.

unique compared to the other sites' structures because it is much larger, was built with more care, and consists of a partition with stone slabs forming two chambers with six rectangular niches built into the interior wall.[28] Each chamber had a central hearth and a stone cist burial covered by a triangular stone located just within the interior of the doorway so anyone entering the building had to pass over the burial cist. It has been suggested that the building was a link to the ancestors and possibly served as a ceremonial building for the Barnhouse residents. Furthermore, each house at Barnhouse was systematically torn down at the end of its occupation, but Structure 2 was left intact. Does the husked barley found in Structure 2 suggest that this was also a beer-production site? Structure 8 at Barnhouse, which sits opposite Structure 2, was constructed in the shape of a large hall with thick outer walls, was built on a platform of yellow clay, and had an internal courtyard that extended more than 20 meters in length.[29] Before one reached the inner hall and courtyard, one had to walk over a hearth through a three-meter-long passageway. The hearth may have served to purify someone before entering the ceremonial structure. Structure 8 did possess large Grooved Ware pots that were too large to move and were partially buried into the floor, similar to what I have seen the Gamo do in Ethiopia with their beer pots to keep the pot from tipping over and to keep the beer cool. Many pottery sherds were found in the structure as well, and residue analysis from these sherds indicates barley lipids or fatty acids and "unidentified sugars."[30]

Bronze Age Beaker-Folk and Beer

We know from scientific analysis that Beakers in many parts of western Europe, including Britain, were used for alcoholic drinks such as mead and ale.[31]

Andrew Sherratt[32] wrote that bell beaker vessels were a cultural hallmark during the Bronze Age (3200–600 BCE) and were used for drinking beer. Bell beaker vessels are found throughout Europe, date from 2750 to 1800 BCE, and are usually associated with burials.[33]

The Bronze Age beer-drinking Beaker folk originated in southern Europe and migrated north, eventually replacing the early farmers who built the megalithic structures such as Stonehenge.[34] Beaker pots were placed in front of the decaying Southern Circle at Durrington Walls near Stonehenge.[35] However, the beaker vessels left there did not have evidence of beer but rather dairy residues in the form of fatty acids or lipids found in the walls of the ceramic vessels.[36] Although the beaker vessels at Durrington Walls did not reveal beer, the technology associated with beer production and consumption followed this new form of technology known as bell beaker vessels, which most likely functioned as drinking vessels.[37]

Other Bronze Age sites in the United Kingdom, such as the Yarnton and the Forteviot sites, have revealed evidence of possible beer.[38] The Yarnton site situated along the upper River Thames has charred and ground barley grains, suggesting either bread[39] or beer production during the late Neolithic/early Bronze Age Periods (c. 2900–1400 BCE).[40] The Yarnton site was contemporaneous and occupied the same region of Britain as the Stonehenge-Durrington site complex, spanning from the Neolithic (3,600 cal BCE) to the end of the Bronze Age (1400 BCE).[41] The site has multiple features, from houses, middens, enclosures, to a cemetery containing cremated remains.[42]

The Scottish site of Forteviot has one of 80 henge monuments located in Scotland.[43] The Forteviot site's evidence of beer comes from residues recovered from a 25-year-old woman who was buried in a large sub-circular cist dating to the Early Bronze Age (2199–1977 BCE).[44] The residues revealed a high percentage of meadowsweet (*Filipendula ulmaria*) pollen and grain (*cerealia*) pollen that may have been a por-ridge or fermented ale.[45] The young woman was buried within the center of the main Bronze Age henge (2468–1938 BCE), suggesting that she may have come from a high-status lineage.[46]

Other Scottish Neolithic and Bronze Age sites have inferential ev-idence of beer production.[47] The stone circle site of Machrie Moor on the Isle of Arran has pottery dating to 3000, BCE with evidence of cereals and honey from residue analysis suggesting that the pots may have been used to store a sweet beer.[48] Pottery known as Grooved Ware dating to c. 3000 to 2000 BCE from Balfarg, Glenrothes, Fife, Scotland, have residues of cereals, meadowsweet, and henbane.[49] Lime and meadowsweet pollen were found in a beaker vessel buried in a Bronze Age cist containing the body of an older man from Ashgrove Farm in Fife.[50] The amount of lime (*Tilia cordata*) was more than 50 percent of the pollen present, and since lime trees are native only to Wales and to southern Britain, the beaker vessel was most likely filled with a mead beer made from lime honey and flavored with meadowsweet flowers.[51] Other Scottish sites with pottery residues that may suggest Bronze Age beer production were found on the island of Rhum off the west coast of Scotland, dating to c. 1940 BCE.[52] The residues indicate evidence of cereal pollen and heather (*Calluna vulgaris*), royal fern (*Osmunda regalis*), and meadowsweet (*Filipendul ulmaria*). It is reported that the Scottish distillery William Grant & Sons successfully reproduced a beer using these ingredients.[53]

The earliest evidence of beer in Spain dates to the Copper Age (c. 2400–2500 BCE), when the living placed bell beaker jars at two ancestral Neolithic sites, La Peña de La Abuela site (3800–3700 BCE) and the La Sima Mound site (3700–3600 BCE and 1860 BCE), both located in the Ambrona Valley in northwest Spain.[54] During the Copper Age, individuals visited the Neolithic tombs and buried their dead with offerings of wheat beer in bell beaker jars. It seems that the Neolithic tombs held some type of reverence and power to the people who were eventually buried there during the Copper Age.[55] Placing the dead with beer some 1500 years after their Neolithic ancestors constructed the monuments was possibly a deliberate act that connected the past to what was the present.[56] Bell beaker vessels are unique because they represent only a small fraction of the entire ceramic assemblage found in Spanish Copper Age settlements. The majority of the bell beaker vessels have a volume of one liter, suggesting that they were suited to drinking.[57] However, some are large, with a volume ranging from two to 10 liters and sometimes more than 20 liters (five gallons), indicating a function for communal storage and consumption, and they exhibit intricate decorative motifs, suggesting that potters spent additional production time producing vessels that they knew would have a special function and be seen in a public setting.[58] These traits, in association with definitive evidence of residues indicating beer from nine bell beaker vessels found from six sites in the Iberian Peninsula,[59] strongly suggest that bell beaker vessels' primary function was for beer consumption in ritual contexts.

One of the earliest and most famous iconic archaeological sites with early evidence of beer in northern Europe comes from the Danish site of Egtved, dating to 1400 BCE.[60] Egtved is a large earthen tumulus with a burial of a 16-to-18-year-old woman with a birch bark bucket containing residues of lime, meadowsweet and white clover pollen, wheat grains, sweet gale, cowberry, and cranberry, most likely used to make a sweet, wheat-based beer. This may have been the same drink that the mourners drank as they worked together to build the earthen tumulus to celebrate the life of this young woman. She was buried in an oak coffin and laid on a cowhide with the hair toward the woman.[61] She was dressed in a beautiful wool corded skirt and wool blouse with a disc-shaped bronze belt symbolizing the sun that may have served as a symbol for a priestess of the Nordic sun religion.[62] She wore around each wrist bronze bangles, and an earring was found on one side of her head.[63] A yarrow flower, found at her knee, is one of the most common medicinal plants in the world for wounds, digestive problems, respiratory infections, and skin issues.[64] Sadly, a cremated child, who was five to six years old, was buried next to her.[65] The child has been assumed to possibly be the young woman's sibling, since she would have been only 13 at the time of birth if she were the mother of the child.[66] The yarrow flower may have been a symbol of how they were trying to treat her, or it could have been simply a symbol of love.

Who was this woman who was buried with beer? Archaeology, specifically the biomolecular and biogeochemical analyses (i.e., tooth enamel, hair, fingernail, and wool fibers) tell us that she came from an area outside of Denmark, most likely the Black Forest of southwestern Germany, and in her final months traveled across large distances eating terrestrial foods with some periods of food stress without protein as she was traveling from place to place.[67] The beer that her friends and family left her and the child were certainly a symbol for them to take on their journey into the afterworld.

The two key traces of beer from the Bronze Age are residues from ceramic vessels placed in association with burials and charred and ground grains. While the grains could be evidence of bread production, many of the sites where these two archaeological signatures of

Figure 5.3 Egtved, Denmark, funerary mound.

possible beer are found contain elaborate monumental architecture that would imply an organized group of people who may have been drinking beer as a form of food in the construction phases and as part of feasts that celebrated both the living and the ancestors. The world's ethnographies support this idea of beer being a major actor in the construction of architectural buildings and monuments from houses, burial chambers, and stone and timber henges. Evidence is growing in other regions and periods of Europe countering old ideas that beer was not an important drink; one of these places is ancient Greece.[68]

Who Said the Ancient Greeks Didn't Drink Beer?

From Neolithic and Bronze Age Greece, it is well established from the archaeological evidence that wine was an important drink from northern to southern Greece. Literary evidence of wine and lack of discussion of beer has strongly suggested that the Greeks did not drink beer but preferred wine.[69] Archaeological signatures of wine production come from grape pips and/or pressings as well as residues from ceramic vessels.[70] However, there is growing archaeological evidence that beer was a common beverage during Neolithic and Bronze Age Greece.[71] Pottery vessels from the Neolithic site of Halai, located on the North Euboean Gulf of Greece and dating to the Middle and Late Neolithic Periods (6000–5300 BCE), indicate severe interior pitting on large-diameter vessels.[72]

Evidence of sprouted cereal grains and cereal fragments from the northern Greek Bronze Age sites of Archondiko (2135–2020 BCE) and Argissa (2100–1700 BCE) suggests that beer production occurred alongside grape pressing for making wine.[73] Both sites consisted of household structures that were destroyed by fire,[74] which may have been caused by fire used to dry the malted grain or heat the wort.[75] The 58 sprouted wheat grains found at Archondiko were not with non-sprouted grains, strongly suggesting that brewers were using these sprouted grains to make malt for beer production.[76] Compared to the sparse amount of sprouted wheat grains found at Archondiko, the amount of sprouted grains uncovered from Argissa is truly remarkable, consisting of at least 3588 sprouted einkorn wheat, emmer wheat, and barley grains. In addition to the sprouted grains found at Archondiko, cereal fragments and lumps were also discovered, suggesting possible evidence of a beer mash that was used as a starter for yeast inoculation for the wort.[77] In association with the archaeobotanical evidence of beer brewing from Archondiko and Argissa were kantharoid-shaped drinking cups with two high handles on either side of the vessel.[78] At Argissa, within the house with the sprouted grains were 45 cups with a volume ranging from 0.1 to 0.3 liter. Thirty kantharoid vessels were found at Archondiko with a range of 0.1 to 1 liter. The high frequency of these vessels and the volume in association with the archaeobotanical remains strongly suggest beer drinking during some type of ritual feast.[79]

Malt-drying kilns and scenic composition on pottery from the Early-Middle Bronze Age site of Kissonerga-Skalia in Cyprus (c. 2300–1650 BCE) indicate the presence of brewers. The drying kiln was used to dry malt or to cure malt cakes for beer production.[80] The presence of ground stone tools and pottery vessels, associated with the drying ovens, is also a strong indicator of beer production.[81] The ground stone and hand stone tools may have helped the brewers to grind the malted grains into cakes for later use, and the locally produced, large, wide-mouth cooking vessels may have been ideally suited for mash tuns. Other evidence includes a high concentration of jugs with spouts that may have been used as an additive, such as a sweetener or a herbal flavor, to the beer.[82] An intriguing clue is the high frequency of broken and mended bowls found near the drying ovens that brewers might have used to dry the malted grains.[83] A truly remarkable indicator of beer production and feasting at Kissonerga-Skalia is a series of scenic compositions on pottery vessels.[84] These compositional scenes of appliqué art appear on large bowls, jugs, and jars. On 10 vessels, there are indications of beer production, especially with scenes of a pouring trough, individuals or couples drinking from a bowl or jug, a vessel into which liquid is being poured from a trough, and a decorative horizontal and vertical relief band that may have symbolized drinking straws.[85]

New and exciting research is now challenging the strongly held belief that ancient Greeks did not consume beer. Recent research documents beer production and consumption by Greek communities beginning in the Neolithic and continuing into the Bronze Age.

Iron Age Brewery and the Taste of Celtic Beer

Celtic culture is associated with banquets and feasts through a number of sources from Greek and Roman writings, legal documents from Ireland, epic tales from both Ireland and Wales, and lastly from the archaeological record.[86] During the beginning of the Celtic Period, only the elites traded among the Mediterranean societies and chiefdoms to the north in Germany and France. The feasts of ale were reciprocal between the Celtic chiefs and his nobles, whom the chief needed to maintain his strength, sovereignty, and social order.[87]

The early Celtic settlement of Hochdorf, Germany, dating to the fifth to fourth centuries BCE, served as the rural residence of an early

(a)

(b)

Figure 5.4 Kissonerga-Skalia beer bowls showing appliqué.

Celtic prince.[88] Structures identified at the site were a bow-sided house and pit houses, and features included earth cellars, grain storage pits, a fence, postholes linked to storage features, and six ditches. Two of the ditches exposed thousands of sprouted grains of multirow hulled barley (*H. vulgare*) on the bottom of the ditches, used to make malt as part of the Hochdorf brewery.[89] The ditches may have functioned as a drying kiln that had dried mud bricks on the top with a wooden frame that may have held a cover made from woven reed, willow, or textile to dry the malted barley.[90] Based on the archaeological reconstruction of the drying kiln, it caught fire and the grains, drying material, and frame collapsed and fell into the ditch. This ill-fated event for the brewers and consumers was a gain for history because it luckily preserved this moment in time for future archaeologists to use to unravel the mysteries of early beer production.

So what did Celtic beer taste like? Hans-Peter Sitka[91] tried to re-create Celtic beer using interpretations from his excavations at Hochdorf, Germany. Similarly with Viking beers, only the archaeological record can provide us with information on Celtic beer, since there are no written or iconographic sources. The quality of the Hochdorf malt was of high quality when using modern malt production as a benchmark. The Hochdorf kiln had an open fire, most likely adding a smoky flavor to the malt. In addition, the kiln may have unevenly heated the malted grain, which may have produced a dark malt and, because the drying process was slow, may have caused a high rate of lactic acid bacteria. Pottery containers for brewing were not found at Hochdorf, so Sitka[92] infers that brewers used wooden vessels to mash the dissolved malt into a pulp. Cooking stones may have been used to heat the mash to boost its sugar content and caused a caramelization on the surface of the cooking stones, giving it a unique flavor similar to a steinbier. The possible smoky flavor that the Hochdorf brewery produced is similar to modern German beers, such as Rauchbier,[93] that are so common with breweries found in Bamberg, Germany.[94]

Archaeobotanical evidence in the form of germinated grain for malting comes from the Iron Age (500 BCE–1000 CE for Scandinavia) settlement Uppåkra in southern Sweden. Uppåkra is the largest Iron Age site in southern Sweden, containing hall buildings, ceremonial houses, and workshop areas as well as places dedicated to grain storage and ritual ceremonies. The germinated grain was found adjacent to a kiln structure but was absent from the houses, suggesting that the kiln area was a specific brewing area at the settlement. The kilns at Uppåkra were being utilized to germinate grains into malt for beer for at least a 400-year period (c. 300–700 CE), suggesting a specialized craft-brewing area that was most likely engaged in large-scale beer production for the purpose of ritual feasting and/or trade.[95]

Other sites in Scandinavia have revealed germinated malt grains, but they are rare finds.[96] Store malt found in two ceramic pots in a burnt-down house was found at the Danish site of Osterbolle, dating to the first century CE.[97] At the ringfort site of Eketorp, archaeologists have uncovered what they believe to be malt in the form of germinated barley dating to the sixth century CE,[98] as well as several liters of germinated grain found in a ninth-century CE burnt-down Viking Age hall-building.[99]

Before the beginning of the Anglo-Saxon Period, the Romans occupied much of Europe, including Britain, where brewers supplied beer to the Roman army during their occupation around 100 CE.[100] The Roman author and historian Pliny the Elder (23–79 CE) described beer production in Britain, France, and Spain, with the Spanish beer preserving for several years.[101] Others, such as Tacitus (56–120 CE), described beer as "horrible," but Celtic communities continued to brew their beer, particularly in northern areas of the Roman Empire, such as present-day Britain, Belgium, and Germany.[102] After the Romans retreated, there was a great revival of beer in Anglo-Saxon communities, originating in the German region of Europe and extending into Britain.[103]

Anglo-Saxon Beer, Butter, and Funerary Urns

The Anglo-Saxon Period in Britain was a time of transition as the Romans moved out and left a military vacuum in southern Britain for the Scots (in what is now Ireland) and the Picts (tattooed people who lived in what is now Scotland) to move south and invade southern Britain.[104] Against this threat, the British sought help from the Romans, but the Romans were unable or unmotivated to come to their assistance. The British then asked the German Anglo-Saxon tribes to step in as mercenaries to protect them from the northern tribes.

By the time the Anglo-Saxon Period began in Britain, starting in the mid-sixth century CE, beer was a common drink even among small children.[105] Legend had it that the Romans introduced drinking houses, but after their retreat, the Anglo-Saxon mercenaries burned them all down and it took a couple of centuries for alehouses (*eala-hus*) and inns (*cumen-hus*) to become established.[106] Certain domestic brewers became known for producing high-quality brew, and when a new batch of beer was ready to be drunk, the family would tie a branch to a pole, indicating that the neighbors could come over for a drink. This was the beginning of alehouses that would extend into the Middle Ages.[107] Household brewers, known as "ale-wives," were producing most of the beer during this time, but the demand for ale was extensive, since it was part of every meal and ritual feast.[108] One of the stronger ales produced in these alehouses was known as bride-ale, brewed for religious ceremonies. It was so strong that it was blamed for unruly, violent feasts and regulated by local officials.[109]

Highly decorative funerary urns are common among Anglo-Saxon burials, with some of the urns adorned with humanoid or zoomorphic eyes that appear to be looking in all directions. Some pots were stamped or incised with heads and beards, using single- and double-sided bone combs.[110] The combs also have eyes incised into them and were sometimes placed within the cremated burial as an offering. Nugent and Williams[111] argue that the Anglo-Saxon funerary urns with design motifs consisting of eyes may have been meant to allow the dead to continue to view the living and the afterworld thereafter. Some researchers believe that potters produced the wares specifically as a depository vessel for the newly deceased's ashes and the pots had no association with beer production or consumption.[112]

However, at the site of Cleatham cemetery (fifth/sixth centuries CE), located in southeast Britain, Perry[113] found evidence of beer consumption by analyzing these pots with a new perspective based on use-alteration analysis. Perry[114] observed a variety of use-alteration attributes, from abrasion on the vessel's base to interior pitting, that are consistent with my ethnoarchaeological research indicating that the Ethiopian vessels were used for storing beer. Of the 958 urns from the Cleatham cemetery, 265 have interior pitting, strongly suggesting that these vessels were used for storing beer and were not produced just for storing a cremated individual. Interestingly, around 10 percent of the urns from Anglo-Saxon cemeteries have post-firing holes punctured into the bases and lower walls that may have served to separate the liquid from the solid portion of the beer or butter.[115] Brewing and drinking of beers would continue with the Viking Period. and the Vikings would carry their brewing knowledge as far as Iceland.[116]

Viking Beer as Food

The ancestors of the Vikings liked their beer sweet, as the archaeological residues from pots indicate a grog-like beer made from honey, local fruits such as bog cranberry and lingonberry, and grains or wheat, rye, and/or barley.[117] Trade from southern Europe in the form of wine was sometimes added to the grog, indicating that there were culinary interactions between northern and southern Europeans. These beers were being produced throughout Denmark and southern Sweden around 1500 BCE.

Vikings living in Scandinavia during the eighth, ninth, and tenth centuries CE are often associated with images of seafaring conquerors, but what was their daily life like?[118] Although beer is not discussed in written or iconographical sources of the Vikings until the Middle Ages (c. 1200 CE),[119] archaeological evidence indicates that they produced and consumed beer and used wild hops and bog myrtle to flavor and preserve their beers.[120] Both ingredients have been identified from the eighth-century Danish site of Ribe, and bog myrtle has been found in a number of Danish sites.

One of the primary foods that Vikings brought with them as they settled in Iceland in the ninth century CE is barley beer.[121] Large-scale feasts were a socio-political hallmark of Viking culture, where chiefs would provide conspicuous consumption of foods, including beer. The feasts were a way to create alliances and compete with rival chiefs for the support of followers in other territories.[122] However, in Iceland the growing season for barley is shorter than in the Norse ancestral homes of Scandinavia.[123] Historical texts indicate that Icelanders could not produce enough barley for their demand for beer, and by 1194 CE the Icelandic Norse began to import barley to the island.[124] Icelandic sites such as Hrísbrú (tenth–eleventh centuries CE), where archaeologists found one of the largest Viking Age longhouses and texts, indicate that it was occupied by prominent Viking leaders and was the site of feasts/[125] The longhouse has evidence of dense concentrations of barley seeds consistent with beer production and consumption.

All Hopped Up around the World

Before hops became a staple in European beers, brewers from the Neolithic into the Middle Ages employed additives such as meadowsweet, sage, juniper, and bog myrtle for their excellent preservative and flavoring qualities.[126] While hops are not an essential ingredient for the majority of Indigenous beers around the world, they have become one of the most important flavoring and preservative agents in ancient and modern European beer styles.[127] Wild hops are found throughout the northern latitudes (i.e., 35–55 degrees north latitude) from Asia, Europe, Siberia, and North America.[128] Hops' natural habitats are marshy or wet hallows found in waterlogged wooded areas such as alder and oak woods.[129]

Unfortunately, it is not known when European brewers first used hops in their beer.[130] The domestication of hops in Europe may have begun during the early Roman era (c. 500 BCE to 50 CE) based on pollen records.[131] However, based on written records, hop cultivation and eventual domestication did not begin until the ninth century CE and was fueled by the need to preserve the beer for longer periods of time.[132] The earliest record of hops is disputed, with some arguing that hopped beer was being produced by ancient Finns in 1000 BCE based on a Finnish epic poem *Kalevala* describing beer with hops.[133] However, this claim is disputed by those arguing that the oral tradition story was not published until 1835, the poem's meaning most likely changed over this long time span, and it was most likely influenced significantly by the introduction of Christianity to Finland in the eleventh century.[134]

Another early mention of hops in Europe is from 768 CE when Pepin le Bref donated *Humlonarias* from the forest of Iveline to the St Denis monastery.[135] What does the name *Humlonarias* mean? The common interpretation by previous researchers is that it meant "hop gardens," but Wilson[136] contends that *Humlonarias* is a place-name. In addition, there is no record of hops in Pepin's son Charlemagne's detailed list of plants within the estate.[137]

However, the first written evidence of wild hops comes from the *Statutae Abbatiae,* dating to 822 CE and written by Abbot Adalhard.[138]

After a tithe of the hops has gone to the monastery, part of the supply of hops should be given to him [the porter] every month. If, however, this is not enough of him, then he should be buying or in some other way acquire a supply (of hops) so that he has enough to brew beer.[139]

This quote was made some three centuries before what most researchers state as the earliest record of hops.[140] The quote from Abbot Adalhard suggests that hops were not cultivated yet but were gathered from the forests, indicating that the demand was low enough around 822 CE that brewers relied on the gathering of wild hops from the woods.[141] Most historians had placed the earliest evidence of hops at ca. 1150 CE with Hildegard of Bingen, who had the foresight to discuss the preservative qualities of hops based on their bitterness.

Before the cultivation of hops for the primary purpose of brewing beer, there were many different uses of hops related to health and other purposes, such as to clean the blood of "impurities, tumors and flatulence" and "relieve the liver and spleen."[142] Hops were known to have antibiotic substances that helped to curtail certain bacteria, such as those causing tuberculosis.[143] Medieval apothecaries often added their remedies to beer for their patrons.[144] Besides for medical purposes, there were many other uses of hops: ancient Greeks used hops for salads,[145] and hop salads are still part of the culinary tradition in parts of central Europe.[146] In the mid-nineteenth century, Californians used hops as a household ornamental plant.[147] In Sweden, hop stems were woven into fabric that resembled linen cloth. Hop extract in Russia was used to dye brunette hair.[148]

The earliest evidence of cultivated hops comes from Bavaria, dating from 859 to 875 CE and involving hop gardens in the Abby of Freisingen.[149] The 40 or so years between the time of the earliest record of wild hops (822 CE) and the eventual hop cultivations (c. 860 CE) indicates that the demand grew enough for monasteries to spend the labor and time to cultivate hops.

The cultivation of hops for brewing did not occur at the same time throughout Europe.[150] Britain had hopped beer in the tenth century CE,[151] but brewers did not add hops on a regular basis until the beginning of the sixteenth century CE.[152] After the sixteenth century, Britain, Germany, and the Netherlands brought hops with them as they expanded their colonial empires.[153] For example, growing hops in North America began in New England and New Netherlands in the seventeenth century and in Virginia in 1648.[154] British colonists brought hops to Australia in 1788 as well as into Tasmania and New Zealand.[155] German brewers brought hops to China and Korea in the 1860s and brought the German and American varieties of hops to Japan in 1876 rather than using the Japanese varieties. The history of hops documents that while it was not at first in high demand and had many different uses it is now almost solely used to brew beer.[156]

Witchcraft, Ergotism, and Beer in the Middle Ages

While beer has been and continues to be an essential food for many societies both in the past and present, some brewers may have used contaminated grain, which may have caused substantial harm to the consumer. One of these is the catastrophic disease called ergotism, caused by ingesting the fungus *Claviceps purpurea* occurring in contaminated rye used in beer or other types of foods.[157] Ergotism during the Middle Ages was widespread,[158] with a reported 83 epidemics of ergotism throughout Europe beginning in 945 CE[159] that killed many people, but the cause was not yet understood.[160] Rye (*Secale cereal*) is the leading grain that became contaminated with the fungus that causes a variety of symptoms, including gangrene, convulsions, hallucinations, headaches, nausea, diarrhea, formication (i.e., the feeling that insects are crawling on you), and difficulty speaking. Rye grows well in wet and cool areas with acidic soils. It was the last crop to be introduced into Europe after barley, oats, and wheat and became more common during the pre–Roman Iron Age in central Europe and Russia. It became really popular following the mass migrations after the fall of the western Roman Empire and into the early medieval period.

Witchcraft, ergotism, and beer were interconnected, with at least 23 witchcraft trials involving beer or milk used as a means of causing harm to others.[161] In the Norwegian 1663 witchcraft trial of Kristina Skryppa, Aksel Skryppa, and Ragnhild Myklebust, the accused claim to have learned the craft of causing harm by inducing the beer with a hallucinogenic substance, most likely ergot.[162] Kristina and Aksel stated they learned the "art" of witchcraft by drinking beer. When Kristina drank the beer made by another brewer, the inoculated beer made her hallucinate and the devil came to her. The devil presented himself to Kristina in the form of a "goat kid" and called himself "the flying bird." Aksel Skryppa was given beer by his mother, which had him reject his baptism and Christianity, and after he drank the beer the devil appeared as a black, spotted cat. Both Kristina and Akset attended a black Sabbath at Hornelen Mountain on the west coast of Norway.[163] Other Norwegian examples of using beer to induce witchcraft come from the case of Gunnhild Oldsdatter, who was tried and found guilty of witchcraft in 1663. She says she learned witchcraft from a woman who gave her beer.[164] This beer contained a "magical substance," and then Satan took control of her.[165] This "magical substance" was described as "blue starch," which she was told to place in the food of the schoolmaster's children, but instead she gave it to a dog that went crazy, ran into the sea, and drowned.[166] Most likely, the brewers used ergot-infected rye malt.[167] Dorethe Larsdatter was condemned to death in 1662 because she gave beer to a man who after four days died a terrible death.[168] Another death sentence was handed out to Karen Monsdatter, who stated during her trial that she learned the art of witchcraft from Narve-Ane, who gave her beer, and then she was able to fly wherever she wished.[169]

European Beers: Variety Is the Spice of Life

Development of global beer culture and modern beer styles are rooted in specific, historical brewing centers around the world.[170]

The Bavarian Purity Law (Reinheitsgebot) of 1516 restricted brewers from using any other ingredients other than water, yeast, hops, and barley malt for bottom-fermented beers, and for top-fermented beers brewers must use the same ingredients but could use wheat malt as well.[171] Other ingredients for special beers were allowed, including beet-cane and coloring agents originating from sugar. During the fifteenth and sixteenth centuries, wheat was not used in Bavaria because it was considered too valuable for beer and was only allocated to making bread. However, in much of ancient Europe (c. 3000 BCE–1000 CE), wheat beers were the most common type of beer.[172] Beer produced for the poor was made using cheaper ingredients such as oat malt rather than the more expensive barley malt and cheap bittering agents such as willow bark and ox bladders instead of hops. Wheat was eventually used to make beer when Bavarian royalty and wealthy merchants began to produce beer, but the Bavarian Purity Law continued in earnest in Bavaria because it was considered a food or "liquid bread" with an average per capita consumption in Bavaria of 240 liters (63.4 gallons).[173] Although the purity law allows for only four basic ingredients, German brewers were able to produce a variety of beers, from bocks to lagers to pilsners.

Classic European beer styles such as ales, lagers, lambics, porters, and stouts have their own histories, and many, such as Kölsch, Lambic, and Pilsner, have names based on their location of origin.[174] European beers fall into either an "ale" or a "lager" category.[175] The word "ale" is thought to have originated from the Anglo-Saxon word "Ealu", whereas "lager" is derived from a German word meaning "to store," first used in 1420.[176] Ales are based on top-fermenting yeasts occurring in warmer brewing temperatures, resulting in a fruitier fragrance and fuller body,[177] while bottom-fermenting lagers use yeasts that react to colder temperatures and need to age longer to permit volatiles such as diacetyl and sulfur to dissipate.[178] The chemistry of water has also influenced different types of beer styles and where they originated[179]. For examples, pilsners use soft water that is common in the Czech Republic, and pale ales originated in Burton-on-Trent, Staffordshire, England, because of its hard water high in calcium and magnesium.

Trappist Beer—480 CE

Trappist beer originated in medieval Normandy, France, but it is now brewed in only eight monasteries.[180] Trappist beer consists of different styles; therefore, it is a category all to itself.[181] It began to appear during the Middle Ages, brewed by a variety of religious orders, but St Benedict (480–547 CE) is thought to be the originator of Trappist beer. One of the hallmarks of the Trappist monasteries instituted by St Benedict was being self-reliant, inspired by Jesus' wanderings in the desert,[182] where he fasted for 40 days and 40 nights and was tempted by the devil for food and power. The tenet by which the monks brew their beer is that all the beer must be produced within the wall of the monastery by the monks or under their supervision.[183] Beer is a secondary endeavor, and all profits from the production of beer go to supporting the monks and their abbey as well as running the brewery and caring for the community. They would also give beer to pilgrims and travelers and would trade beer with the community to maintain the abbey, since the monks could not produce many of the needed items. The variety of Trappist beers include Single and Triple, which are usually blond in color and consist of complex aromas, including cloves, fruit (usually apple or pear), and bubble-gum![184] The Dubbels and Quadrupels are a brownish-red color and are maltier in taste, similar to darker breads and dark fruit, such as figs, prunes, plums, and dates, and have a sweetness to them because of their high sugar content.

Lambic Beers—1559 CE

Lambic beers are unique and represent their own beer family, neither an ale nor a lager.[185] They are known from the Pojotten land region just southwest of Brussels, Belgium, and have a deep history, not only in

Europe but also in many regions of the world.[186] The earliest documentation of Lambic beers as a brewing style dates to 1559, thought to signify the village of Lambeek 12 miles southwest of Brussels.[187] A royal decree that went into effect on May 20, 1965, defined and protected beers that were spontaneously fermented.[188] Unique to lambic brewing is the transporting of hot wort into cool vessels or open tuns called "cool ships," which were located in the attic of the brewery.[189] This period, called "pitching," is when the airborne yeast and additional microbial flora inoculates the wort.[190] This pitching only occurs from October to April when outdoor temperatures around Brussels are under 15 degrees Celsius. Then the brewers put the inoculated wort into oak or chestnut casks that had been previously used to ferment wine, since new casks would cause too much tannin to enter the lambic beer. Before the beer is placed in the casks, the brewers scrape and clean the interior of the casks and burn a sulfur wick in them to kill any detrimental molds. Belgian law states that to be considered a lambic beer. the beer must have a minimum of 30 percent malted wheat.[191] Lambics are aged in wooden port barrels from a minimum of six months to six years. There are a variety of lambic beers, including Lambic, Gueuze, Biere de Mars, Faro, and Fruit Lambics.

Porters and Stouts—1700s CE

How porters and stouts began is still in debate, but the common belief is that brewers created what is known as "three threads" consisting of equal proportions of ale, beer, and two penny (a strong beer costing twopenny, or tuppence, a quart).[192] Ralph Harwood was a brewer in Shoreditch, London, and he brewed a re-creation of three threads using roasted malts and called the beer Entire." It was a popular beer among the porters at the shipyard and was named after them. Porters became the first mass- produced beer during the Industrial Revolution and grew internationally.[193] Baltic porters developed with the expansion of British colonialism, but they had a higher alcohol percentage and were more hoppy so that they would not spoil during the long voyage to the Baltic nations.[194] In England, the public wanted stronger porters, so brewers developed stouts, which eventually replaced porters. This was the beginning of Arthur Guinness's stout as well as Imperial Stouts shipped to Russia and the Baltic countries and Foreign Extra Stouts shipped to the West Indies.[195] Stouts use black patent barley that has a dry, intense flavor, whereas porters are made with roasted malts that have a chocolate and coffee flavor. Recently, Guinness Brewery was forced to stop using isinglass, which is a gelatin from the air bladders of freshwater fish such as sturgeon, to clarify the beer by separating out the yeast and the solid particles.[196] Guinness began to use isinglass in the mid- to late nineteenth century, and it does not change the taste of the beer but helps the settling process occur more quickly. Vegans were turned off by the use of isinglass in Guinness and began drinking German and Belgian beers that were regulated by the Bavarian purity laws.[197]

Pale Ales and India Pale Ales (IPAs)—1700s CE

As with the invention of refrigeration for pilsners, the use of coal for heating allowed for the development of pale ales by having controlled heat to malt grains gently and creating a lighter-colored malt.[198] Burton-on-Trent, England, was the heart of pale ale brewing with light-colored malts; brewers could produce different flavors of pale ales with caramel, toffee, and bread flavors. This was very appealing compared to the dark roasts of porters and stouts, and with the advent of an orange- and amber-colored beer, pale ale caught on and became a new popular style of beer that would span the world. One of the largest colonial British outposts was India, but the local alcohol drink was *arak*, a spirit that did not provide the needed nutrients of carbohydrates, amino acids, etc. that the colonists required. British soldiers and other colonial actors would need beer in their outposts. Mark Hodgson, a London brewer from the Bow Brewery, decided to take the pale ale and modify the recipe so that the beer could travel the long 6000-mile voyage from Britain to India.[199] Hodgson increased the hop content, and he cleverly put dry hops in the barrels of the finished beer to increase the chances of preserving the beer.[200] Hodgson's innovation also helped to stabilize the beer against the constant rocking of the boat. The added hop flavor was a perfect thirst quencher to compensate for India's hot and humid tropical climate. In 1835, the ship *Stirling Castle* shipwrecked off the north coast of Australia, with the crew, of eight out of 19 cast ashore and alive on Cumberland Island with only three gallons of brandy and an 18-gallon barrel of Hodgen's pale ale.[201] Besides drinking the beer, they were able to chew on the spent hops at the bottom of the barrel.[202] Hodgen continued to export pale ale to India and controlled the trade with India using the British East India Company until 1821.[203]

Kölsch beer—1800s CE

Kölsch beer was a blond and lightly hopped ale brewed by Cologne (Köln) brewers.[204] The earliest trade organization, the Guild of Brewers, began in 1396, it has enforced the unique style of Kölsch beer even up to this day. In the 1800s, as part of a reaction to the influx of bottom-fermenting golden pilsners into the Koln area from Bohemia and Bavaria, Kölsch brewers began to brew a golden Kölsch beer made with pale malts, locally grown hops, and top-fermenting ale yeasts; that is, it was made with cool fermentation but was longer aging.[205] It is reported that Kölsch leaders did not permit pilsners to be distributed in the town of Kölsch.

Bohemian Pilsners—1842 CE

Bohemian pilsners began to be brewed around 1842 in Pilzen, western Bohemia (currently in the Czech Republic), and were the first golden, clear beers in the world.[206] This would have a profound impact on the types of beers people would begin to drink, not only in Europe but also around the world, because people were fascinated by this new style of beer. Prior to this, beers throughout Europe and the rest of the world were cloudy and dark because they were unfiltered. Gabriel Sedlmayr of the Spaten Brewery and Anton Dreher, a Viennese brewer, were vanguards in developing brewing methods using pale malts from England and Belgium. Spurring the popularity of lager beers was Carl von Linde's momentous invention of refrigeration using an ammonia compressor first installed in a Trieste, Austria, (Trieste was part of Austria at this time and later was annexed by Italy), brewery in 1876.[207] Amazingly, this refrigerator did not need to be serviced for 10 years after being installed and lasted 32 years! Burger Brauerei, well known today as Pilsner Urquell, was the first brewery to brew a pilsner in the Czech Republic. The golden color of Bohemian pilsners that were complex in taste, including a sweet and spicy flavor but with a low bitterness and rounded off by its soft, mineral-free water, made pilsners popular and became the foundation for some of the best-selling beers today, e.g., Budweiser.[208]

Sour Beers—Ancient Times to Today

The Indigenous beers, past and present, are sour beers, and the drinkers of sour beers today should thank the world's ancient brewers, who created an array of aromatic sour beers. German and Belgian brewers created Berliner Weisse, Belgium's Oud Bruin, Flanders Red, and lambic beers.[209] Eventually, European immigrants would bring to America their unique local brews, such as the sour Berliner Weisse.[210] Unfortunately, because of the US government's 14-year prohibition on alcohol, sour beers production did not survive as well as lagers and ales.[211] Several European brewers began to import European sour beers, but it was not until the mid-1990s that American craft brewers were inspired by the world's diversity of beer tastes and began to brew sour beers.[212] Kinney Baughman of Cottonwood Grille and Brewery in Boone, North Carolina, was one of the first to commercially produce a Belgian Amber Framboise and Black Framboise, in 1995. They mixed a clean beer with a brew that had been accidentally inoculated with the lactic acid bacterium *Pediococcus*.[213] Then New Belgium brewery followed with a sour beer, La Folie, that is considered as good as any Belgian sour beer. After La Folie won first place at the 2002 Great American Beer Festival, other brewers, such as Russian River Brewing Company, followed New Belgium's lead. The popularity of sour beers has skyrocketed, with almost every craft brewery producing a sour beer, usually aged in wine and spirit wooden barrels and seasoned with local fruits and spices.[214]

Conclusion

From the colossal stones and timbers of Stonehenge and Durrington Walls, respectively, it seems likely that beer helped to motivate workers to move, carve, and erect these giant boulders.[215] Once the people had erected the stones and timbers, they feasted, and we hope that in the near future archaeologists will find concrete evidence that beer was on the menu. Other contemporaneous sites in the UK have strong signs of feasting, and by 2000 BCE the ritualized bell beaker drinking vessels were common throughout Europe.[216]

New research documents that beer was a common drink during the Greek Neolithic and Bronze Ages,[217] dispelling the widely held perspective that Greeks enjoyed only wine. Around the time the Greeks were drinking beer, Scandinavians were also drinking beer and building large funerary mounds.[218] Beer throughout different regions of Europe was tied to rituals, with drinking vessels associated with burials during the Anglo-Saxon period in Britain[219] to the Middle Ages, when contaminated beer caused abnormal behavior attributed to witchcraft.[220] Eventually, wild and then domesticated hops became an important additive for preserving and flavoring beer.[221] Over the past five centuries, European beer styles were adopted and have now become popular around the world. Some of these varieties, such as Belgian lambic beers and French Trappist beers, were being brewed back in the Middle Ages, but other beer styles, including pilsners, stouts, and pale ales, have a more recent history dating to the past two centuries.[222]

We now move across the Atlantic to the Americas, where from northern Mexico to the tip of South America societies were brewing corn beer, known as chicha beer. This beer was and continues to be one of the most important foods that motivated workers to build some of the most iconic archaeological sites and gave symbolic majesty to their sacred spaces.

6

Meso- and South America

Beer Fuels Runners, Roads, and Feasts

The antiquity of beer in Meso- and South America may be as ancient as in Southwest and East Asia.[1] While there is some indication that beer production may have begun very early, the bulk of the evidence of brewing beer comes from later state-level societies, especially in South America.[2] There is strong evidence that ancient societies brewed beer from northern Mexico (c. 1200 CE) to the tip of South America (c. 1200 BCE) as a medium to connect the living to their ancestors and celestial sprits.[3] While prominent societies such as the Maya and Aztec Empires relied on other forms of psychoactive drinks made from chocolate or balché, a tree bark, for their ritual use,[4] chicha beer has a long history of ritual, economic, and social importance to Meso- and South American societies.[5] Most chicha beer is brewed from maize, but many other ingredients can be added to maize or substituted as the primary fermentable ingredient (see Chapter 2).[6] Contemporary Indigenous women continue to brew chicha as a daily and ritual food and to unite individuals and lineages during important community projects.[7]

This chapter discusses how chicha beer has been one of the most important daily and ceremonial beverages.[8] From the present-day Raramari in northern Mexico and Quechua-speakers of the Andes, to the ancient Peruvian polities of Chavin, to the Inca, beer coalesces each of these communities.

During the sequence of ancient Andean polities, beer is the medium that connects the spirit world to the living, manifesting in the monumental construction of ritual centers.[9] The workforce to build these conduits to the spirit world must have been fueled by the consumption of chicha. Commissioned by the elite, stonemasons carved elaborate channels for the purpose of feeding beer to their sun god and their ancestors.[10] The elites connected to their deities and ancestors by administering to these ritual monuments, bringing about harmony and fertility through the consumption of chicha. Consuming chicha and feeding these powerful celestial spirits, such as the sun god, was the belief in calming the chaos that perpetually exists on our planet. This reciprocal act of feeding deities and ancestors allowed them to aid the living by bringing fertility to the crops, livestock, people, and to all things, living or dead.[11]

Beer in Meso-America

Stalks before Grains

Maize or corn (*Zea mays* spp. *mays*) is the primary grain for beer production in Meso- and South America. Around the same time that Old World farmers were engaging in cultivating and domesticating their barley and wheat fields, New World farmers were also planting squash, maize, and, eventually, beans.[12] Although maize eventually became the dominant crop in the Americas during its initial cultivation and domestication, it remained a minor crop until c. 1000 BCE.[13] Unfortunately, how, when, and where maize began to be processed for beer production continues to elude archaeologists. The earliest wild maize, teosinte (*Zea mays* ssp. *parviglumis*), and domesticated maize had small cobs and small kernels but had plenty of sweet stalks to turn into a fermented beer.[14] This has led some researchers to suggest that the wild form of maize, teosinte, may have been domesticated for its sugary stalk to produce beer rather than its grain.[15]

The Stalk-Sugar Hypothesis argues that Early Archaic peoples in Mexico may have produced beer by casually harvesting the domesticated sweet stalks of maize and chewing them, and later increased production by mashing and squeezing the stalks.[16] The hypothesis is supported by remains of maize stalks and quids from cave sites located in the Tehuácan Valley of central Mexico dating from 3500 BCE to 1540 CE.[17] Another series of caves in northeastern Mexico just south of the Texas border revealed 151 chewed maize quids.[18] The frequency of chewed maize stalks decreased over time in accordance with the rise

Figure 6.1 Map of Meso- and South America—Map showing sites and societies discussed in the chapter.

of maize as a food source.[19] Maize stalks also have been uncovered in Peruvian cave sites, but none have evidence of being chewed.[20]

If brewers were processing maize stalks by mashing and squeezing, then high concentrations of phytoliths (microscopic plant silica) may be present in the soil or on ground stones.[21] Deciphering maize phytoliths that come either from stalks or cobs could help to determine if brewers were processing the maize for beer or for other types of food.[22] Dental wear from masticating maize stalks is another possible archaeological signature of processing maize stalks for beer production.[23] Beer production from maize also may be discovered through ground stone and ceramic residue and use-alteration analyses, which are methods that we have seen in the previous chapters as one of the ubiquitous analyses for deciphering ancient beer.

Casas Grandes/Paquimé, Mexico

When the Spanish first visited the site in the 1500s, they likened its beauty to a Roman city, filled with colorful paved alleyways and beautifully painted mosaics on walls rising three or more stories tall.[24]

The above quote describes the splendor and wonder of Paquimé, a site located in the present-day Mexican state of Chihuahua that may have been the northern-most region of beer production in the Americas.[25] It is here at the site of Paquimé that archaeologists have found the only direct evidence of ancient beer-making in northern Mexico. Beginning around 700 CE, the Indigenous people began to build the regional center of Paquimé (700–1450 CE).[26] The Casas Grandes cultural influence extended from northern Mexico into the southern portion of the American Southwest.[27] The site of Paquimé was the economic and ritual center of the Casas Grandes region and contained apartment-like puddled adobe walls some 50 cm thick for homes as well as platform mounds for astronomical viewing, ball courts, ritual structures, a sophisticated municipal water system consisting of reservoirs, drains, wells, and canals, breeding areas for macaws and turkeys, and millions of shell offerings most likely brought in from the Gulf of California.[28] It is considered a major pilgrimage site for the region, bolstering the authority of elites.[29] Beer production at Paquimé and its surrounding sites is recorded archaeologically from pitting on ceramic vessels and from analysis of human dental calculus (i.e., plaque).[30]

Pottery vessels from four communities tied to Paquimé reveal the interior pitting witnessed in so many other sites worldwide, indicating

Figure 6.2 Paquime site.

that beer production from corn was a common food in northern Mexico at this time.[31] The pottery assemblages from the four sites indicate variation in their interior pitting. In the domestic context of two small sites (sites 231 and 217), the ceramic assemblages had little evidence of pitting. However, the two large sites (sites 204 and 242) with ball courts of the classic I shape and evidence of macaw breeding have a high frequency of pitting on the interior of the ceramic assemblages. Site 242 possibly served as an administrative site for the region; it was built with thick walls resembling the architecture found at Paquimé, and interior pitting was found on large vessels, possibly for feasting.[32] Experimental research with corn fermentation does indicate that fermentation causes pottery vessels to exhibit pitting.[33]

A second line of archaeological evidence for beer at Paquimé and its hinterlands is the analysis of calculus or plaque from teeth.[34] Calculus analysis is a new and exciting method of understanding what a person ate days to weeks before death.[35] When a person ingests food, part of the food materials adhere to the teeth as plaque and can be scraped off after death to reveal microfossils attached to the tartar.[36] Burials from Paquimé exhibited evidence of calculus showing evidence of maize starch granules that had been damaged because of fermentation.[37] An earlier dated site of Convento (700–1200 CE) did not indicate any evidence of modified starch from calculus, suggesting that brewers began to make beer in the later period when Paquimé was being built (1200–1450 CE).[38] However, further research testing this theory will need to be undertaken. Calculus analysis, in conjunction with pottery use-alteration analysis, is a promising method that has the potential to significantly add to our deciphering of beer consumption in the ancient past.

Beer Feasts and Runners in the Sierra Madres of Mexico Today

I was met with incredulous laughter when I suggested to informants that it might be possible for some wealthier men to make a continuous supply of tesgüino [beer] for their own private use. It therefore does not produce the bizarre pathology of Western alcoholism.[39]

One of the most popular contemporary societies that produce and drink beer on an almost daily basis are the Rarámari, who live in the Sierra Madre of northern Mexico.[40] They have been popularized by Christopher McDougall's[41] best-selling book *Born to Run*, which documents their use of beer to fuel their long-distance running. The Rarámari's word for beer is *tesgüino*, made from sprouted maize, referred to as chicha in Spanish. One of the unique aspects of the Rarámari is that they will travel by foot, sometimes running, through the Sierra Madre Mountains when they hear that someone is providing a beer feast. The distance that they run is staggering, running up to three hours through difficult mountainous terrain. For the Rarámari, beer permeates their entire society from their rituals to their economics, health, and technology.[42]

The Rarámari consume their beer as a social activity rather than as an individual act.[43] Similar to other beers brewed worldwide, the Rarámari beer does not have a long storage period. The high expense and the short storage period are two reasons that the beer must be finished at a beer feast. When beer is drunk at gatherings, it is dedicated to *onoruame*, the sprit who gave the Rarámari the knowledge of brewing. The beer feasts/*tesguinadas* can last as long as 48 hours. *Onoruame* gave the Rarámari beer "so that they could get their work done and that they might enjoy themselves."[44] Each gourd of beer must be first dedicated to *onoruame* through symbolically feeding the spirit with beer by pouring beer in each of the four cardinal directions. They feed *onoruame* first so he does not become angry. After *onoruame* symbolically drinks the beer, the beer sponsor presents the beer to the most influential elder, who then serves the beer in order of social rank. Beer provides rain and protects and cures all living things, including crops, animals, and people. Ritual specialists protect each newborn baby by dipping their crosses into the beer and placing the cross on the newborn, who is then fed a small amount of beer. Beer helps to heal all members of Rarámari community, and if a person is ill then the doctor and patient will drink a small amount of beer. As in the case of the newborn infant, the doctor will place his cross, which has been dipped in beer, on the wrists and head of the patient.

Apart from curing purposes, cooperative work parties are one of the primary reasons for brewing *tesquino* beer.[45] As with other Indigenous farming societies, the Rarámari prefer to conduct tedious work such as weeding, planting, harvesting, fence-making, or house-building with cooperative labor, and beer is part of the payment for communal work. The beer acts as a binding force among individuals, families, and communities and reinforces the social and economic obligations and reciprocity that cooperative work instills. When the host invites other members to attend a "working *tesquinada*," he says, "Would you like to drink a little *tesquino* tomorrow?"[46] He describes the work that will be involved and then states that there will be a *tesquinada* to follow. There is no one pattern to the structure of the *tesquinada*, with some household members doing the bulk of the work before the other workers attend, and vice versa.

Beer is directly tied to Rarámari wealth and status, similar to the Gamo of Ethiopia.[47] Wealthy Rarámari have the grain surplus to produce beer, even several months before the harvest time. A person's status is tantamount to the distance people will travel to attend a wealthy individual's beer feast, indicating that a person's social and economic network is a sign of his status. Beer is a symbol of social

control, and if you are left off the guest list, then your social status is
reduced.

Beer in South America

The Roar of the Chavín Pilgrimage Site, Peru

The production of maize beer in South America is largely supported
by research in Peru, beginning with the Chavín pilgrimage center (c.
1300 to 600 BCE) at its ritual site of Chavín de Huántar.[48] The magnificent spiritual center of the Chavín Period was the monumental site of
Chavín de Huántar, which was built in a sacred space located between
two rivers on the valley floor of the Mosna Valley at 3150 masl, considered to have been spiritually powerful for rituals.[49] The site also is
situated near a snow-capped high peak and natural hot springs, both
of which are associated with the worship of deities.[50] The exterior
sunken plazas would have isolated ritual participants from the outside
world.[51] The massive monumental stone architecture is engraved with
art representing humans, animals (such as jaguars), plants, and images
combining all three of these images and reflecting anthropomorphic
supernatural or masked and costumed priests.[52] The priests had feath-
ered headdresses and clothes, jaguar pelts, a monkey spirit playing a
Strombus shell trumpet, and other spirits holding stalks of the hallucinogenic mescaline cactus San Pedro (*Trichocereus pachanoi*).[53] One
possibility is that participants who attended the ritual feasts at Chavín
de Huántar were mixing the psychotropic plant San Pedro with chicha
beer.[54] In 1653, the Spanish Father Bernabe Cobo described how the
Inca would summon "the devil" by mixing their chicha beer with
the juice of the psychotropic juice from the vilca plant (legume tree
Anadenanthera sp.) or ayahuasca (*Banistreriopsis caapi*).[55] Besides
mixing different hallucinogenic drugs with beer, they may have sniffed
the crushed drugs as a form of snuff.[56].

It is possible that the psychotropic chicha beer was consumed to
enhance the spiritual experience of walking through the galleries at
Chavín de Huantar. The site is built within a stone-faced mound with
interior stone-line drained canals and labyrinth galleries.[57] These

galleries may have been built for the purpose of running water through them to cre2ate a roar that would be associated with the animal spirits inhabiting the sacred space.[58] One of the most famous features of Chavín de Huántar is the "oracle" gallery of the Lanzón, where a 4.5-meter-high granite shaft was carved into the form of the Lanzón.[59] The Lanzón sculpture sits partially sunk into the floor, and the ceiling in the center of a small, squarish gallery depicts a half-human–half-jaguar figure with fingers, toes, and claws, human ears decorated with earspools, but a jaguar-like face that exhibits "tusk-like canines" with eyebrows and hair carved as snakes.[60] Pilgrims hallucinating from ingesting ayahuasca would have entered the Lanzón gallery, and as they approached the Lanzón, they would not have been able to view the entire stone but would have viewed portions of the carved creatures looking back at them. The hidden chamber located directly above the Lanzón may have served as a place where ritual leaders talked with or answered the questions of the members who were visiting the Lanzón. If a pilgrim was hallucinating, they may have seen the Lanzón move, speak, and blink its eyes.

The site contains caches of broken bottles and bowls, elaborately engraved shell trumpets, and offerings of exotic materials associated with ritual ceremonies.[61] Within Chavín de Huántar underground chambers there were hundreds of ceramic bottles and bowls, indicating evidence of possible beer consumption.[62] Individuals brought most of the bottles and bowls from Peru's south, central, and northern regions to make an offering to the sacred site of Chavín de Huántar.[63] One bottle was decorated with a maize ear with its husk partly peeled back, suggesting strongly that chicha beer was being drunk during these ritual feasts.[64] Future research needs to focus on the use alteration of these ceramic bottles and bowls to determine what substances (i.e., chicha and/or hallucinogenic plants) were being consumed at Chavín de Huántar.[65] Chavín de Huántar, with its elaborate sacred symbolism that depicted diverse images, would attract people in different regions to make a pilgrimage to this sacred site to feast on chicha beer and other foods and leave offerings to their spirits.[66] Later, a new polity would emerge along the Peruvian coast, with beer as one of the primary foods helping to motivate laborers to construct monumental buildings.

Figure 6.3 Chavín de Huántar underground chambers.

Sacrifice in Moche, Peru

The Moche polity (200–850 CE) relied on a combination of rich coastal resources and intensive agricultural crops on the northern valleys of coastal Peru.[67] The Moche are known for the production of chicha beer as well as for their beautiful stirrup-spout ceramic vessels, advanced metallurgy, and ceramic vessels portraying conquests and human sacrifices.[68] They built their capital at Cerro Blanco, with their two massive monumental administrative structures, Huaca del Sol and Huaca de la Luna, standing 500 meters apart.[69]

At the Moche site of Pampa Grande in context with the ramped platforms, there were jars decorated with faces used to prepare and drink chicha beer.[70] The platforms at Pampa Grande were the places where rites were performed involving chicha production and consumption.[71] In association with the many decorated beer vessels at Pampa Grande were clay flutes, figurines, and talismans, giving us a picture of musicians as part of the ritual scene while participants were drinking their chicha.[72]

Figure 6.4 Chavín de Huántar Lanzón.

Face-neck jars similar to the ones found at Pampa Grande were also uncovered at ceremonial sites in the lower Jequetepeque Valley.[73] The vessels portray Moche religious thoughts with fanged deities, wrinkled faces, and warriors. The Moche's ontology or view of their world and its relationship with beer can be clearly seen through the archaeology of Huaca Colorada (600–850 CE), one of the monumental sites located in the Jequetepeque Valley.[74] The builders of Huaca Colorada constructed a monumental elongated platform mound measuring 390 meters by 140 meters and rising up to 20 meters, where the elite lived.

These elite leaders were the ones who were associated with elaborate feasting jars containing chicha beer and showing images of Moche religion and cosmology.[75] The series of platforms found at the site are where the leaders displayed their drinking of chicha beer to the others who lived and visited the huaca.[76] Adjacent to the platform was a well-preserved chamber with stucco columns that most likely supported a gabled roof and where the elite walked in organized processions and feasted on foods, including maize beer.[77] Feasting was not the only activity occurring within the chamber; there rested evidence of the sacrifice of two adolescent women, one with a rope tied around her neck, with a guinea pig, a dog, and a concentration of crushed copper prills (the remelting of metal into a new form).

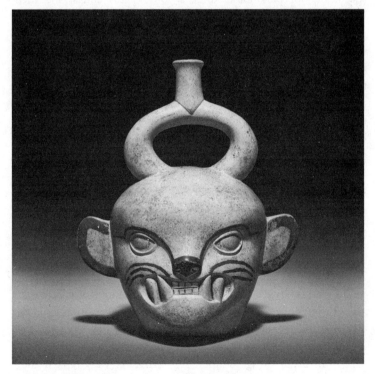

Figure 6.5 Moche bottle vessel with feline head.

The harvesting of maize, which symbolizes its sacrifice for the brewing of beer in association with human and dog sacrificing with the copper prills, supports the Moche worldview or ontology in which death must occur for life to continue.[78] The harvesting of maize and then brewing it for beer may have acted as a symbol of renewal.[79] Many of the offerings, including the skull of a sea lion, etchings of warrior figures, and litter-bearers carrying decapitated human heads, are found in the liminal/transitional spaces of blocked doorways, signifying this transformation between life and death.[80] These platforms would then be covered over as an act of death, while a new platform would be created and brought to life.

The Moche coastal polity had an organized group of high-ranking male religious specialists (*chicheros*), who brewed the beer for ritual feasts, most likely with the help of women brewers.[81] They brewed specifically for paramount *kurakas*, lower-status chiefs, and their followers.[82] While women did brew and consume beer, it was tied to men and their religious authority. The feasting jars are decorated with male faces, suggesting to Andean researchers that high-ranking men were engaged in competitive feasting associated with warfare during this time.[83]

The elite's demand for chicha beer as part of their ritual feasts was at the expense of the rural laborers who toiled as part of the *mit'a* system in which married adult males provided several months of labor for the state every year.[84] The inequity led to uprisings and the destruction of Moche symbols. Eventually, the first empire to construct a sacred site with a brewery producing a unique style of chicha beer would appear.[85]

Wari Empire and Beer in the Sky

The first Andean empire to appear in the central Andes was known as the Wari (600–1100 CE), who conquered and influenced communities throughout the region.[86] They created alliances throughout the diverse ethnic regions by sponsoring rituals and commensal feasts that included chicha beer to legitimate their political control and establish debt from neighboring communities.[87] Hierarchically ranked elites controlled the Wari Empire, centralized at their regional capital in

Ayacucho, Peru, and at conquered hinterlands in western and southern Peru.[88] High in the Andes is the remarkable site of Cerro Baúl (600–1000 CE), located in the southern frontier of the Wari Empire, where on top of this sacred 600-meter promontory mountain brewers constructed the largest known ancient brewery in the Andes.[89]

Cerro Baúl was built for its defense, sanctity, and politics.[90] The site, located in Peru, was built during the Wari Empire, which ruled over a 1,300-kilometer/808-mile territory of the Andean mountains.[91] The Wari's reign was well before the Inca took control of the Andes, which did not come about until some 400 years later. For the current Indigenous community, the mountain is sacred (*apu*), meaning that it is "revered as earthly spirits that protect, but may also punish their human constituents. These *apu* were often linked to distant ancestors and are considered the most important local deities."[92] The sacred site of Cerro Baúl, both ancient and modern, was an unlikely place for anyone to have built monumental architecture, including a monumental brewery, palace, D-shaped temples, and administrative and storage buildings.[93] Furthermore, all the building materials, including the water for the chicha beer, had to be hauled up the steep slope to the site. However, here we have a spiritual relationship between Indigenous landscapes and the production of one of the most important foods for the Andean people, chicha.[94] *Apu* are fed chicha beer as a sacrament to maintain the health and fertility of crops, animals, and people. Therefore, the construction of Cerro Baúl's brewery makes perfect sense.

The brewery is unique in that it contains not only the brewing technology but also vestiges of the brewers themselves and the last elite drinkers of chicha.[95] The brewery had four distinct rooms for grinding maize, preparing food, boiling the mash, and fermenting and storing the chicha.[96] The milling room had a thatched roof, and the floor contained high levels of phosphate from the milled corn, which the brewers would grind on the ground stones.[97] The boiling room contained *tupu* shawl pins, strongly suggesting that elite women were preparing the chicha beer. It is believed that the women brewers would undo their pins to remove their *tupu* shawls because of the heat from boiling the mash. The boiling room contained between eight and 12 hearths, each with a pair of opposed stone pedestals, thick ash deposits,

Figure 6.6 Photo of Cerro Baúl, prominent position in the landscape.

and a high frequency of seeds of *Schinus molle*, which is a spicy berry that brewers boil and soak whole to release sugars in the resin pockets on their central pits.[98] The brewer discarded the pits and boiled the syrupy mash that would make the fermented chicha de molle. After boiling, the brewers moved the chicha de molle to the fermentation area containing 12 large ceramic vats that lined the north wall of the central patio.[99] Each fermentation vessel was decorated with a human face on the neck of the jar and a chevron design representing the head-band painted on the rim.[100] Here the chicha fermented for three to five days in vats that could hold up to 150 liters of beer, indicating a production capacity of around 1,800 liters per batch, the largest pre-Inca brewery to be discovered in the Americas!

Cerro Baúl seems to have been systematically shut down permanently in conjunction with the last production of chicha de molle beer.[101] When the last brew was ready to be drunk, it was served from ceramic pitchers to 28 nobles, who drank from beautifully painted vessels, or *keros*, with images of deities painted on the sides, the vessels

Figure 6.7 Photo of the Cerro Baúl brewery's boiling room showing where the large vats of beer were placed.

ranging in size and status from 12 to 64 ounces. After the drinking, the brewery was set ablaze and ritually burned. As the brewery burned, the nobles threw their vessels into the fire in an act of sacrifice and eventually placed six necklaces of shell and stone and a bracelet atop the ashes in a final act of reverence.[102] By the time the Wari Empire started to wane, the Chimu and the Incas began to ascend, sometime after 1000 CE, with the Inca becoming one of the most powerful Indigenous societies in the Americas, if not the world.[103]

Chimu's Beer Economy: Peru

Chicha was politically important, a beverage that could be turned into labor, and this was true for Andean polities other than the Inka Empire.[104]

Succeeding the Moche state along the northern coast of Peru was the Chimú Empire (900–1470 CE) during which the lords used chicha to pay their corveé ,workers.[105] The monumental capital of the Chimú Empire was Chan Chan, which was architecturally unique, with its builders using adobe and walls reaching more than nine meters (27 feet) in height. The walls of the monumental buildings have sea creatures etched into the adobe walls that became popular at the end of the Moche Period.[106] The Chimú Empire, similar to previous Andean empires, was extremely stratified, with the majority of the population composed of craftspeople, including brewers, as well as families engaged in farming, fishing, and shellfish collecting.[107] The Chimú kings lived in Chan Chan and built ciudadelas that were tied to a specific ruler. The ciudadelas were palaces with open courtyards, kitchens containing dense cooking materials, and multiple storerooms, some ciudadelas having more than 200 storerooms. The residence areas contained walk-in wells with the audiencias that served for working quarters for the king's family, palaces, burial platform, and plaza where large beer feasts would occur.[108] The craftspeople and other laborers lived in densely occupied houses that were made from cane wattle and mud daub with cobblestone foundations.[109]

Similar to other ancient regions of the world, the Chimú State relied on laborers to conduct public work for the state, such as canal and road building, copper work, woodworking, farming, and fishing. In exchange, the Chimu provided beer feasts.[110] Of all the crafts, chicha production, as the above quote by Moore[111] states, was political. This

Figure 6.8 Photo of the adobe walls of Chan Chan.

was apparent when the Spanish colonial official Gonzalez de Cuenca instituted a ban on distributing chicha along the north coast of Peru in 1566 (just after the Inca state ended), which thereby halted all work, since the community leaders could not compensate their laborers with beer.[112] After five days, the order to begin redistributing beer was reinstated after Cuenca realized that beer was the political glue that created the motivation for the completion of public works.[113]

As the Chimú state gained power over time, they moved north to the Jequetepeque Valley, located 100 kilometers north of the Chimú capital of Chan Chan, and conquered this region (1250–1450 CE).[114] The production and drinking of beer played a prominent role in the ritualized life in this valley, with formalized architectural construction for ritual feasts for connecting the dead ancestors to the living.[115] The rituals encompassed drinking chicha beer with dancing, playing music, offering ritual items, and the eventual burying of elite ancestors.[116] A miniature wooden model of these ritual beer feasts was unearthed from a Chimú burial ground at the monumental site of Huaca de la Luna, providing us with a rare glimpse into the rituals.[117] Within the wooden model are three groups of musicians who are serving chicha beer, with the *chichero* (server of beer) ladling the beer from the fermenting jar into the drinking jars.[118] The archaeological record collaborates on this ritual feasting with evidence of marine shell, ash, abundance of broken cooking vessels, and other organic materials.[119]

When the Chimú leaders took control over the Jequetepeque Valley, they increased their chicha production to fuel state-controlled irrigation projects that would lead to higher agricultural yields in the valley.[120] Evidence of beer production from rural households at the Chimú site of Pedregal indicates an increase over time of large fermenting vessels for chicha production, quantities of maize dregs from the spent wort, and large hearths for heating the beer.[121] Chicha would have been an important daily food for the people living in the Pedregal community but would have also been distributed to administrative centers for ritual feasts and supported large-scale state-managed projects.[122]

Brewers at San José de Moro (1350–1470 CE) in the Jequetepeque Valley produced large quantities of beer, as discussed in Chapter 2, which was possibly produced for the master brewer living at Huaca Alta and for the elite at the nearby administrative center of El Algarrobal de Moro.[123] The master brewer who lived at San José de Moro was from the elite class, and the skill of turning maize into chicha most likely was believed to be an act of ritual change and performance.[124] Ritual offerings were made to the brewery, including young llamas, dogs, guinea pigs, textiles, small ceramic vessels, and marine shells.[125] One of the more intriguing offerings was the placement of textiles and a large wooden paddle over the fermentation jars found in the Area 35 brewery.[126] The large wooden paddles found at Area 35 are similar to an early Chimú ceramic bottle that depicts a master brewer holding a large paddle. A brewer's paddle also is depicted in an 18th-century watercolor of an official Chichero/master brewer holding a paddle in front of the Chichero's tomb.[127] The offerings and especially the wooden paddle were important symbols of an elite brewer and were possibly presented to the ancestral spirits during times of architectural change or as part of a ritual transformation occurring within the community.[128]

Figure 6.9 Photo of the large wooden paddle over the fermentation jars found in the Area 35 brewery at the site of El Moro.

Inca's Dead Kings, Roads, and Beer Capital

Beer was not necessarily the secret ingredient that enabled the Inca to establish the largest Native empire of the New World, and one of the largest in history, but it certainly was one of a series of interrelated key features that enabled the Cuzco rulers to extend and maintain their power over a vast region. . . .[129]

Beer feasts began with the Chavín and continued with the Moche, Recuay, Tiwanaku, Wanka, Wari, and Chimu polities before the Inca Empire took control of the Andes.[130] The Inca Empire (1350–1532 CE), known by its Indigenous descendants as Tawantinsuyu (land of the four quarters), was one of the most powerful ancient states in the world.[131] Chicha beer (*aqa* in Quenchua) was intertwined with their religion, economy, technology, and health.[132] Maize, the main ingredient in chicha beer, was not just a simple food resource but represented a symbol of Inca ideology and identity.[133] The Incas established a social and economic infrastructure that extended from present-day Ecuador to central Chile with 25,000 miles of roads.[134] Their capital, Cuzco, was known as *aqa mama*,[135] meaning "mother beer," and stonemasons built their capital with high-quality stone blocks that could weigh more than 100 metric tons and fit so tight that a razor blade could not pass between the two colossal pieces of stone.[136] These stones were sacred to the Inca, meaning they were *huaca*, and one of the items that the Inca offered these sacred stones was chicha beer.[137] Rock carvings at the outcropping site of Kenko Grande, located just outside of Cuzco, were constructed for the purpose of pouring liquid libations, such as chicha beer, down a channel (*paqcha*) designed as a zigzag-carved pattern.[138] This movement of beer through the rocks symbolized the connection between the upper- and underworlds.

Inca Feasts

> *One must fill up with drink so the extra can be transferred to the ancestors.*[139]

Chicha beer represents a luxury food in the Andes, with its association with ancestor feasts.[140] Inka feasts lasted several days, with participants consuming more chicha than food.[141] While brewing was not a full-time specialization, there were large breweries where the Inca fermented and stored their beer.[142] Chicha could be made from a variety of plants, including oca (*Oxalis tuberosa*), manioc (*Manihot esculenta*), quinoa (*Chenopodium quinoa*), molle tree fruit (*Schinus molle*), algarroba (*Prosopis pallida*), strawberry (*Fragaria* spp.), peanut (*Arachis hypogaea*), pineapple (*Ananas comosus*), and most commonly maize, and if the Inca wanted to make a special chicha, they would bring different varieties of maize from long distances away.[143] Chicha was commonly drunk on a daily basis as well as consumed at large-scale ritual feasts.[144]

Throughout the Inca world, rulers sponsored chicha-centric feasts like the Festival of the Sun (*Inti Raymi*).[145] By serving vast amounts of beer and other foods, the emperor could solidify his power over the population by making them in debt to his generosity.[146] The state had

Figure 6.10 Photo of the sacred rock of Kenko, where libations such as chicha beer were offered.

specialized farmers who grew most of the maize for the ritual feasts, and the surplus maize and other crops were stored in state-sponsored centralized storage sites.[147] These state workers, who were dedicated to growing the best crops for the Inca rulers, also were closely tied to the sun deity that was tied to the elite.[148] The young girls (*acllakuna*) who harvested the grain were taken by the state from their homes at the age of 10 to work either in Cuzco or at one of the administrative centers.[149] Their primary work obligations were to brew beer and to weave textiles, both of which were of paramount importance to the Inca rulers.[150] At one of the sites, more than 200 *acllakuna* were living together[151] and could have brewed enough beer to feed more than 1000 guests at one of the state-sponsored ritual feasts.[152]

The fertility of the Andes and the knowledge of brewing by women brewers (*mamakunas*), whose labor was their tribute to the empire, allowed the emperor to brew millions of liters of chicha beer a year.[153] The Inca emperors viewed these women brewers as holy and considered their beer sacred.[154] The feasts would occur at the capital, Cuzco, and other provincial Inca centers.[155] It is estimated that the amount consumed at some of these feasts was a staggering nine liters per person.[156] The reason for the copious drinking may be an idea that drinking would make the earth happy and fertile.[157] What is incredible is that the Inca elite could not stockpile the beer for future feasts, since the shelf life of beer was less than a week. Therefore, they needed to have state-sponsored brewers (*acllacona*) and to rely on every community lineage to provide beer for these large ritual feasts.

Beer was one of three main foods given to workers as part of their tribute, the other two being meat and maize stew.[158] The Inca redistribution system transformed perishable foods, such as beer, into the construction of temples, palaces, roads, canals, specialized craft items, and the production of crops.[159] Beer vessels were the most frequent artifact found in the Inca hinterlands, indicating that chicha allowed the Inca to expand their empire.[160] Beer was such a critical part of the Inca society that Inca administrators organized state-sponsored pottery production areas where potters were located along the north-south coastal road so that they could move their wares throughout the Inca Empire.[161] Some potters were relocated by the state and attached to provincial centers to produce the Inca flared-rim jars for beer production and serving. One of the reasons the Inca were able to control such a vast territory was the tribute system, whereby conquered craftspeople had to travel and work for the state, and whenever they stopped along the road system, there was beer available to drink.[162]

The economy throughout the different Andean polities relied on compensating workers as part of their tribute labor (*mita*) with chicha beer and food.[163] One of the monumental feats achieved by the Inca State was the road system that linked the empire along the Andes from

Ecuador to Chile and connected key administrative centers and way stations (*tampus*).[164] Constructing the vast road system and its infrastructure, such as the *tampus*, took a large amount of *mit'a* labor from rural communities, and these laborers needed to be compensated with chicha beer.[165] Besides feeding the laborers beer, sacred sites in the form of rocks were revered and given offerings of chicha beer.

Inca Feeding the Ancestral Stones

> *The principal offering, the best and most important part of Indian sacrifices is chicha. By it and with it the festivals of the huacas begin. It is everything.*[166]

Beer and other liquids, including blood, urine, and possibly human semen, were symbolically "cultural water" that brought life to the living Inca by renewing the energy of the sun and the ancestors.[167] For many Inca, receiving life (i.e., *sami* or *enquaa*) comes from the flow of beer and other liquids to stones, people, and places.[168] Achieving the order and harmony as part of *sami* is commensurate with acting responsibly.[169] The proper movement of beer from the gods to the living and the dead would help one achieve harmony in one's life.[170]

The Inca relied on their ancestors to replenish harmony and well-being for the living. The ritual intersections between beer, water, mountains, and the sun may be similar to the modern Quechan belief that the sun moves underground along the Vilconata River during the night and is refreshed by the chicha offerings that allow it to be born again each morning.[171] During the summer solstice, the sun is dim due to the river being low and there not being enough water for the living. Therefore, beer and/or water was poured down these channels during the summer solstice to revitalize the sun. The Quechan perspective of *camay* is where fluids such as beer, water, and/or blood generates "life force, animate[s] material objects, and render[s] them efficacious."[172] Pouring a libation such as chicha beer down a channel gives vitality to all things living and confers sacred powers on individuals.[173] *Usnus* (ritual platforms) and *huacas* (ceremonial shrines) are sacred places where the libations afford this potent life force.[174]

Specific places in the form of symbolized rocks built as platforms and shrines were administered to by pouring chicha on them.[175] Some of these platforms and shrines were modified to include carved channels where the Inca offered chicha beer to *Inti* (the sun), which served as the emperor's deified ancestor.[176] Some *huacas* and *ushnus* were constructed to track the movement of the sun and moon and were built to honor the ancestors.[177] Offering beer to the ancestors was done to seek assistance in agricultural fertility, victories in war, health, and fertility to all living things.[178]

Sacred sites of Isla del Sol, Machu Picchu, and Cuzco had *huacas* and *ushnus* that were both important spaces where ritual participants made offerings in the form of beer.[179] Many of the *huacas* and *ushnus* also had a celestial connection between the underworld, the earth, and the cosmos. The *huacas* shrines included sculpted rocks and ceremonial centers such as Chavín de Huantar, as discussed earlier in the chapter, which was the first sacred pilgrimage space associated with shamanic powers and the offering of beer. *Ushnus* are raised platforms with carved underground channels with drains to absorb the poured beer and other liquids (i.e., water, blood) as an offering to bring harmony to the world.[180]

Numerous women brewers (*mamakunas*) resided at the Sanctuary of the Isla del Sol, which was a difficult pilgrimage site to visit during the Inca time because of its remote location on the northern end of Lake Titicaca.[181] The Sacred Rock (*Titicala*) was the most important sacred place, possibly covered in gold, silver, and a finely woven cloth.[182] The sacred chicha would be poured into a stone basin located at the base of the Sacred Rock, and then the beer continued through a hole at the bottom of the basin to run down a series of stone channels. This area of the site was restricted to the priests an1d the Inca rulers and their attendants, and lower-status individuals could only watch the ritual from afar, some 400 meters away.[183]

The famous mountain-op site of Machu Picchu has more than 30 religious features, many of which are *huacas* and *ishnus* where beer would have been a central libation[184]. Machu Pichu served as a royal estate for the Inca leader Pachacuti, where feasting and hunting were primary activities.[185] Pilgrims would have walked the Inca trail and would have passed through the Gate of the Sun (*Intipunku*) and a security gate before entering the sacred site.[186] Neighboring sites, such as Llactapata, also have evidence of offering beer libations. Llactapata is only five km from the famous site of Machu Picchu.[187] Here in front of the doorway of the Sun Temple in the sacred plaza is a stone-lined channel that runs toward the Sacred Plaza, where chicha and/or water were poured as a libation to the sun.[188] Other locations where beer was poured down a stone-lined canal as an offering to the ancestors were at the *usnu* in the main plaza of Cuzco.[189]

Cuzco was the Inca capital, and one of the names for the Cuzco was *akha mama*, or chicha mother, since *akha mama* is the fermenting agent that is saved and used to start a new batch of chicha.[190] Chicha was used in the Festival of the Sun, Inti Raymi, in planting, and in offerings to the ancestors. Cuzco contained as many as 16 solar pillars built surrounding the capital, where at the sunrise of the summer solstice, known as the feast day Init Raymi, the Inca emperor and his relatives watched the rising sun from the plaza at Haucaypata and drank chicha beer.[191] They then poured chicha or water from Lake Titicaca into a basin that flowed down a rock carved channel to the House of the Sun.[192] Other sites with rock carved channels include stone of Saihuite and the sanctuary of Isla del Sol.[193] At Saihuite, located approximately 160 km west of Cuzco, there are an intricate number of ritual channels oriented to the summer solstice sunrise.[194] One of the channels constructed in a myriad of grooves is associated with carvings of humans, pumas, llamas, frogs, monkeys and lizards. Besides feeding beer to the living *huacas* and *ushnus* to bring harmony to the world, the Inca also fed their ancestors, especially their deceased emperors, during specific times of the year based on the solar and lunar calendars.[195] According to Pedro Pizarro's[196] account, the mummies of emperors were set in a row in the plaza in order of when they served as emperors.[197] Fires were set in front of them to prepare the foods that were to be served and large vessels of beer. The chicha was then poured into the ushnu, a gold-covered round stone with a basin that drained into the earth. The round stone and basin were decorated with multi-colored feathers and an image of the sun.[198] Afterward, large amounts of chicha were brought into the plaza to be drunk by everyone, including the representatives serving all the nations that the Inca had subjugated.[199] This deep connection between beer and religious, social, and economic order and harmony of the world continues to the present day.[200]

Chicha Beer Today

As with other regions of the world, industrial beers and an increase in locally made liquors have reduced chicha production and consumption today.[201] The production and consumption of beer in South America began to decline when viral diseases spread across the Andes following Pizarro's landing on the shoreline of Peru in 1532 and subsequently ordering his army to destroy the sacred landmarks of the Andean people.[202] Production changed after the introduction of cane sugar to enhance the alcohol content of chicha, and other grain crops such as barley, oats, and wheat were added to established ingredients in the brewing of new types of chicha beers.[203] During the first half of the twentieth century, there was an orchestrated movement to vilify chicha by associating it with laziness, poverty, and crime.[204] Today, chicha is still made throughout the Andes as a daily and ritual drink.[205]

To make chicha is to make an ayllu.[206]

Among the Sonoqo of Peru, production and drinking of chicha requires an interconnected network of kin known as an *ayllu*.[207] Many families live in high-altitude zones (3200–4000 meters) where they are not able to cultivate maize and must rely on family members living in the lowlands to supply the grain for the production of chicha.[208] Many families live part of the year in isolation in secondary homes while attending to their highland crops such as potatoes and herding sheep and llamas. The production of chicha is one of the most demanding tasks for rural Andean women, and when they do brew chicha it is in great quantity. Since the shelf life of chicha is limited to a few days, women cannot store the brew, and if there is no one to drink it, then there is little incentive for its production.[209] When chicha is made, it takes men to obtain the maize and the fuelwood and kinswomen as part of an organized reciprocal ayllu to work together to prepare a brew for a community activity such as building a bridge or helping to harvest crops.[210] Unfortunately, chicha has been replaced by either trago, a distilled sugarcane mash, or alkul, pure distilled alcohol, leaving behind all of the ayllu kin ties and community-bonding that chicha provided.[211] Drinking beer, as in so many other regions of the Indigenous world, was an organized community engagement that brought people together without the manifestations of alcohol abuse, similar to what Kennedy[212] found with the Rarámari, as discussed earlier in this chapter. Recently, because of systemically poor economic opportunities for many farmers in highland communities, coupled with the reduction of community engagement centered around organized rituals and work parties sponsored by community ayllus, some individuals have begun to drink individually, leading to a rise in alcoholism.[213]

In other regions of the Andes, such as in Cochabamba Valley in highland Bolivia, chicha remains an important economic driver for women.[214] Perlov's[215] ethnographic study of Bolivian chicheras beautifully outlines the social and economic influence that chicheras have in their community. Brewing and distributing chicha for Bolivian women situates them with more economic and social power by giving them autonomy in their life.[216] Similar to what is occurring in Africa (see Chapter 4), beer production by women allows them to have a consistent income even after the cost of expenses, payment to laborers in the form of chicha, and the customary gift of giving one glass of chicha free to her customers.[217] An important aspect of being a chicheras is having economic independence from her husband, a situation that is rare in urban Bolivia, where husbands may take their wives' salary. However, the chicheras can keep her income because it is secret and separate from the household budget. Economic independence turns into a social influence: since customers cannot always pay their bills for chicha, the chicheras can use her influence to have the customer bargain for agricultural labor that will benefit her economically. Chicha enables the chicheras to gain economic and social power, and they turn this into providing a better education for their children. Sadly, because chicha production demands so much labor, a woman cannot do it alone and relies on her daughters to assist in the daily production activities. In order to maintain this power, the work needs to be done by the daughters rather than asking for help from their husbands or outside laborers. However, because of the strong gender roles, it is the sons who most commonly are awarded the higher education opportunities instead of the chicheras' daughters.[218]

Amazonian communities on the eastern side of the Andes rely heavily on the production and distribution of beer for social, economic, and religious cohesion.[219] Beer in the Amazon is made from manioc, which has been a domesticated crop for at least 8500 years.[220] Beer can be either a bonding agent or a means to mediation.[221] In the case of the community of Conambo in eastern Ecuador in the Amazon, people negotiate their political statuses within domestic houses where women serve manioc beer.[222] Within these domestic confines, it is the women who produce and serve the chicha beer in beautifully painted serving bowls. The women serve the beer to the men in order of their political stature and alliance. She serves her husband first and then men who are closely tied to her family, with others being served in order of status. She can use this serving etiquette to demonstrate the seriousness of the negotiations.[223] Beer is then used as a medium to settle disputes and to bring the community together.

"Upi, cai asuara, asua Runa causimi," which means "Drink this manioc beer, manioc beer is the life of Runa people."[224]

Asu gives life (asua causaira cun).[225]

In the Ecuadorian Amazon, Michael Uzendoski[226] worked with the Napo Runa, who produce manioc beer (asua), as the quote above signifies. The Napo Runa strongly associate beer with human reproduction. Manioc beer gives the Napo strength, happiness, and hospitality. In the household, the woman brewer serves asua. and the entire bowl that is offered is drunk by the participants.[227] Men deeply desire their wives to brew asua, which is to be drunk communally with male kin members. At weddings, the groom's family gives beer to the bride's father, which acts to symbolically unite the two families and reduce ceremonial tension.[228] Many women work collectively to produce enough beer to feed the wedding participants, and the women line up to serve each participant their beer.[229] However, instead of drinking the entire bowl, each participant drinks only a few sips, and then they move to the next woman's brew. One cannot refuse beer at a wedding;

otherwise, a guest will have the beer poured over their head, so they must pace themselves in order to make it to the end of the wedding celebration without becoming incapacitated. Serving is commonly associated with sexuality, with women dancing and singing flirtatious songs to men, especially during weddings, and highlighting the desire to gratify his sexual desires. The above quotes link manioc beer to *samai*, a Napo Rapu belief in the "breath" or "soul substance" that all living and many non-living substances contain.[230] Fermenting beer is *pucuna*, meaning "to blow or mature," and *pucuna* is associated with the sexual energy between a man and woman that creates life. Other symbols intersecting life and beer are the large pot for fermenting mash that is symbolically linked to the female womb. As babies are weaned off the mother's breast, they are fed beer, giving them symbolic feminine strength and life. The intense intersection of beer and culture throughout Meso- and South America is vastly contrasted with the lack of beer throughout the Indigenous societies of North America.

Did North American Indigenous People Drink Beer?

There is ample evidence of cultural interaction between ancient Meso-Americans, who had many different types of fermentable drinks, including beer, with societies from the American Southwest.[231] However, based on the present archaeological evidence, the people living in the American Southwest and throughout the rest of North America seem to have not adopted the brewing of beer from corn or other grains.[232] It may be that North Americans brewed beer from endemic grains 4,500 years ago (i.e., chenopodium, *Chenopodium berlandieri*, and marshelder, *Iva annua*).[233] These crops could have been utilized for brewing beer some 2,500 years before maize became an important staple. In the absence of archaeological evidence, it currently seems as if the farthest north beer was brewed is along the present-day border of Mexico along the Gila River of the Akimel and Tohono O'odham Nations in southern Arizona and, as discussed above, at the site of Casas Grandes/Paquimé and its outliers.[234]

North American societies may have been using psychoactive plants for many millennia to enhance their ritual purposes, and as such they did not need to create alcoholic drinks such as beer. Peyote (*Lophophora williamsii*), jimsonweed (*Datura stramonium*), tobacco (*Nicotiana tabacum*), and yaupon holly (*Ilex vomitoria*) were some of the psychoactive plants associated with inducing ritual trances involving shamans.[235] The harvesting of wild peyote from the fields of southwest Texas has a deep history in the Huichol Native American community, where they embarked on long pilgrimages to collect peyote.[236] Recently, Native American Church members ingest peyote as a sacrament that symbolizes the creator's heart to bring spiritual awakening and to heal the ravages of drug and alcohol abuse.[237] Datura or jimsonweed is a hallucinogenic plant when ingested and has been reported at the ancient site of Cahokia outside of St Louis.[238] Tobacco was a substance documented by the ancient Maya[239] and shaman-priests at the northern Mexican site of Paquimé (1250–1450 CE) used to induce hallucinatory visions to unite them with their supernatural deities.[240] In the American Southeast, the "black drink," a strong stimulant made from yaupon holly, is part of the renewal rituals that bring the communities together, similar to how beer is used today and most likely in the past to unify a community.[241] It was also ingested at Cahokia (1050–1250 CE).[242] It is these alternative cultural stimulants that most likely had been ingested for many millennia, and therefore beer was not needed (except at Paquimé) to fulfill these social, economic, and ritual purposes. However, it should be noted that many other regions of the world had alcohol in addition to other stimulants similar to the ones used by North American Native Americans. Therefore, the mystery as to why Indigenous North American societies seemed to have not consumed beer remains. Further research focusing on dental calculus and technology, such as use-alteration and residue analyses on pottery vessels in different regions of North America, may help to find evidence of beer brewing in North America.[243]

Conclusion

The Americas, south of the Mexico/US border, has a long and storied history of brewing beer. While most of the beer is produced from corn in the form of chicha beer, multiple ingredients and types of beers have been produced over the centuries and up to today[244]. One of the earliest beers may not have been produced from the maize kernels but from the sugary maize stalk.[245] The Meso-American society of Paquimé with its massive ball courts, astronomical alignments, macaw and turkey breeding, and intricate water features relied on beer as a psychoactive food.[246]

Farther south, into the spine of the Andes and along the Peruvian coast, a series of South American polities instilled beer with their monumental architecture and as a daily food to supplement workers for their community service,[247] the earliest was the sacred highland site of Chavín de Huántar, with its zoomorphic and anthromorphic carvings, where religious leaders were mixing chicha beer with mescaline to support their ritual feasts.[248] After Chavín, there was a move to the coast, where potters in the shadows of monumental temples produced beautiful beer vessels decorated with images representing their spiritual world.[249] The highland Wari Empire followed the Moche and constructed Cerro Baúl, where the largest known Andean brewery is adjacent to monumental temples.[250] After the demise of the Moche, the coastal Chimú State compensated their laborers with chicha for building an elaborate city with carvings of water creatures.[251] All of these early societies laid the groundwork for the Inca Empire, one of the most powerful early state societies of the world and where chicha beer was one of the most essential commodities fueling the Inca state.[252] Today, even after the devastation from Spanish colonialism, chicha beer continues to be an important culinary tradition that permeates Indigenous societies' religion, economy, health, and technology.[253]

We have spanned the world of beer from the very beginning to the present day. The final chapter summarizes the major themes discussed throughout this book and presents six new successful beer recipes by archaeologists, brewers, and brewing arts students who use ingredients and techniques from the ancient world.

7

Tapped Out

Robert Braidwood, Jonathan Sauer, and other distinguished researchers debated back in 1953 on the role beer played in the domestication of grains. One wonders how surprised they would be by how much we know in 2022 about the importance of beer to ancient and contemporary societies. I believe they would be shocked to find out that brewers were brewing beer in 11,000 BCE, some 4000 years before the domestication of grains.[1] They would also be in wonderment that the earliest beer was not associated with a house and hearth but within a cave intertwined with ritual, death, animals, and technologically innovative vessels. Over the span of the past 70 years since their debate, we know that beer has had a dramatic effect on the innovations of technology, providing a stable, healthy food for millions of people every day, motivating communities to improve their world through daily labor and enhancing their spiritual beliefs.

The story of beer is a chronicle about how we as a species have interacted with each other, created prosperous societies, survived difficult and challenging times, and ended up where we are today. Beer continues to be a critical food source for millions of Indigenous people today, providing a fulfilling and nutritious meal. After water and tea, it is the most consumed beverage in the world and continues to unite the vast majority of communities through daily and ritual life. It is the catalyst for small and large rituals that unite communities. However, because of its societal importance, the distribution of beer can be used to segregate and divide communities by excluding certain individuals from attending community-led rituals and celebrations.

The Four Leitmotifs

Four themes threaded throughout this book are beer's role in introducing new technologies, ensuring health and well-being, imbuing life with ritual and religious connections, and building economic and political statuses. Each of these topics offers unique perspectives connecting the importance of beer to ancient and modern civilizations.

From Stone Bowls to Refrigeration

The advent of beer has usually been in association with the invention of pottery vessels so that the brewer would have something in which to heat and ferment their brew. I have to admit that I did not think that the earliest evidence of beer would be associated with stone mortars located in a cave some 4000 years before the beginning of grain domestication.[2] It would seem that the knowledge of beer and its ability to alter the mind, fill the stomach, and bring people together during ritual and community gatherings was an impetus to continue to improve on this technology. Stone workers carved elaborate decorative stone vessels in the Near East, which may have been allied to the consumption of beer. Eventually, the invention of working wet clay into different shapes and hardening the clay with fire increased the ability to produce large quantities of beer throughout the world. The use of ground stones to manipulate grain after it had malted was equally an important technological achievement. All of these early technologies should not be discounted as easy to manufacture. Finding suitable clays and stone and then manipulating these materials into functional objects is part science and part art. Beer drinking is also tied to technologies related not to the brewing of beer but to the connection of people and their spiritual world. In Jiahu, China, flutes carved from the wing bone of the red-crown crane most likely linked the spiritual world to the living when their minds were under the influence of beer.[3]

One of the oldest technologies associated with beer is the drinking straw. Today, plastic straws have been banned in some parts of the country because they can be extremely harmful to wildlife. While we do not drink industrially brewed beer from straws, drinking straws have been around for as long as brewers have been brewing, and they are still an important way for beer drinkers to consume beer communally. The production and use of drinking straws occurred as part of the earliest beer in ancient Mesopotamia and Egypt, as well as in contemporary societies of the great lakes region of central Africa.[4] Straws have been made with a variety of materials from reed and metal, and some were compound tools with metal or bone strainers attached to the end to keep any sediment from reaching the drinker.[5] Brewers have employed larger strainers made from woven baskets, used among the Ainu of northern Japan and the Gamo of southern Ethiopia.[6] Prior to 1501, European brewers first used straw to strain their wort but then switched to a mash tun with a false bottom to remove the spent grains.[7] South American brewers filtered their beer twice, using different sieved strainers, and changed the way they used their technology to modify the chicha's alcohol content.[8]

One technological creation that brewers utilized more than any other has been the simple ceramic pot. This creative hallmark of potters mining the clay and small particles (i.e., what archaeologists call "temper") to hold the clay together and then shaping, drying, decorating, and eventually firing the pot led to the rapid rise of beer brewing. Brewers were able to boil their wort, ferment the beer, and then serve the beer using a ceramic serving vessel. For some of these beer- drinking vessels, South American Andean potters crafted beautifully decorated stirrup bottles with zoomorphic and anthropomorphic designs.[9] This revolutionary invention also saved brewers valuable time in collecting fuelwood, since the ceramic vessel can heat the beer quickly and maintain a relatively small range of temperature. Not only did ceramic vessels save brewers valuable time and energy in collecting fuelwood, but this also became a more sustainable technology, replacing having to heat stone, hide, or wood vessels using hot rocks, which takes significantly more fuelwood. Indigenous brewers still rely on what may seem like a simple utilitarian tool, but it has had a dramatic influence on beer brewing and other culinary treats.

Brewers have continuously modified their technology to make better beer, especially modifying the temperature fluctuations for bottom-fermenting beers.[10] Belgian and Dutch brewers utilized deep fermenting troughs to protect the beer from the outside air. Brewers also began to use higher-quality kettles to heat their wort and also utilized plumbing to move the water and wort to and from the kettles.[11] One of the biggest technological transformations was the invention of refrigeration by the Austrian Carl von Linde, which was used to keep beer cold beginning in 1876.[12]

The Health Aspects of Beer

Much to the chagrin of beer drinkers around the world, recent medical research on the consumption of alcohol, including beer, states that we should not be drinking at all![13] However, we know that beer provides essential antioxidants, amino acids, folic acid, vitamins, and minerals.[14] Many Indigenous societies expend large amounts of energy working as farmers and artisans, including brewers, requiring these essential nutrients. Rural Indigenous brewers, who are women tasked to brew the beer for daily and ritual consumption, must harvest, remove the husk, and then grind the grain on a ground stone if they do not have a local mechanized mill. Then they need to collect the water from a source, either a river, a lake, or a well, carrying a large ceramic vessel or a plastic jerry can on their back. The brewer also needs to locate, cut, and carry back the fuelwood to heat the mash and wort. Indigenous beers are thick in consistency, similar to a liquefied porridge, satisfying consumers by filling them up and giving them their daily sustenance. Indigenous beers being in essence a lightly fermented porridge, not as filtered as industrial beers makes them more healthful to drink with higher levels of vitamins such as thiamine.[15]

Beer's low percentage of alcohol kills persistent water-borne bacteria found in natural water sources. Water-borne bacteria kill some 525,000 children a year, and there are a staggering 1.7 billion cases of childhood diarrheal disease each year.[16] While water-borne bacterial diseases proportionally affect children the most, they can also cause adults to die, and in 2019 diarrheal diseases were the eighth leading cause of death among the world's population.[17] This is why Indigenous societies begin to feed children beer at an early age--to reduce the chances of water-borne bacteria killing their children.

The earliest association of health and beer is at the site of Jiahu, China, where we have evidence of a grog-like beer made from rice, grapes, honey, or hawthorne fruit.[18] The burial analysis at Jiahu documents a population that was relatively healthy.[19]. In South Asia, the early author Charaka had a perspective on drinking alcohol similar to what we have today, suggesting that as long as people drink moderately and in social groups, it can enhance one's health.[20] Along the lower and middle Nile some 1,600 years ago, people were drinking inoculated tetracycline beer caused by the bacterium streptomycetes that became mixed with warm dough.[21] The burials lack bone lesions, indicating that the beer was a protective therapeutic for these communities.[22] Ancient Egyptians also brewed beer to cure a wide range of health benefits related to mouth and anal diseases.[23] In Europe, the use of hops had a strong connection to specific health benefits, from cleaning the blood to its use as an antibiotic to ward off tuberculosis.[24] In the Americas, the Rarámari use beer as a healing force.[25] Beer did not just affect the biological health of an individual and community. Beer, in both ancient and contemporary societies, brings psychological health as a medium by delivering health, fertility, and harmony from the ancestral spirits to the living.

The Spiritual World of Beer

The advent of beer is spectacularly connected to a ritual cave where a community buried their dead with offerings of flowers and animals.[26] Ritual burials and beer drinking occurred throughout the world in ancient times, often associated with sacred monumental architecture.[27] From the ancient to the contemporary, the production and drinking of Indigenous beer is tied to some form of spiritual belief. This connection between the sacred world and beer is particular to each society, but a common thread since its inception is that among Indigenous societies, beer remains one of the most important mediums linking the living to their ancestral world.

This perspective is found cross-culturally from Southwest Asia, East Asia, Africa, Europe, and the Americas. Beer is one of the foods that the living left the dead as they made their journey into the afterworld. In southwest Asia, one of the largest assortments of drinking vessels is at Gordion, Turkey (700 BCE), the large tumulus where the legendary King Midas may have been buried.[28] The Ainu of Japan believed that the female ancestral spirit *Kamui Fuchi* brought well-being and long life to them when they fed her their spiritual beer.[29] In Gamo Ethiopia, elders pour beer on the living earth to feed the ancestors, who are with them in spirit and bring fertility to all living things.[30] The ancient capital of Great Zimbabwe is still an important sacred site to living people today, and it is here, both in the ancient past and the present, where beer is administered to the ancestors for rain to bring fertility to the crops, livestock, and to all life.[31] In ancient Spain, beer was left with the dead at a site previously used by ancestors during the Neolithic Period, some 1,400 years before the Copper Age (c. 2500 BCE).[32] The large earthen funerary tumulus at Egtved, Denmark, where the living buried a young woman and her possible sibling, contains a birch bark of beer for them to enjoy into the afterlife.[33] During the Incas' reign, beer was fed to the deceased king in a reciprocal act in return for his bringing harmony and fertility to the living.[34] The dead are not the only ones given beer; the living drink beer to inspire and fuel their community work projects.

Feeding the Workers

The Great Pyramids of Giza, Great Zimbabwe's colossal stone buildings, Gordion's and Egtved's earthen tumuli, and the Incas' extensive road system were all built on beer.[35] While other monumental sites, such as Göbekli Tepe and Stonehenge, do not currently have evidence of beer, it would not be surprising if the community leaders were feeding the workers beer as they erected the large stones. Mesopotamia's rich texts document the increased production of beer as cities grew over time, resulting in the building of monumental temples and palaces.[36] The brewery at Hierakonpolis in Egypt could have brewed more than 250 gallons of beer to feed almost 500 workers a day.[37] Another ancient Egyptian brewery, at Tel Edfu, was fueling miners with beer so that they could produce the valuable copper tools for building the Great Pyramids at Giza.[38] From the Peruvian coast to the spine of the Andes, societies built beautiful stone-carved monuments, all fueled by the calories of chicha.[39]

While these monumental landmarks receive all the attention, it is the daily work projects that have occurred over many millennia that should be highlighted in history. These daily work parties that occur throughout the world provide each community with a safety net, allowing not only for cohesion and unity but a means to feed one another. Beer acts as the commodity, allowing wealthy communities or families to transfer their grain and labor wealth into an incentive for community-based projects, such as harvesting crops, building roads, or constructing a new home for a newlywed couple. Beer during the past 13 millennia has positively influenced the technology, religion, economy, and health of Indigenous societies throughout the world. And it has inspired craft beer breweries to learn from these ancient and contemporary brewers.

Coming Full Circle from Indigenous to Craft Beers

In 1980, with only eight craft breweries in the United States, the global beer megacorporations controlled much of the industrial beer production.[40] However, the rapid growth of the craft beer industry over the past 40 years has circled back to the origins of beer, with unique flavors being added by brewers. While the craft beers are different from ancient and contemporary Indigenous beers, the craft beer industry is learning from the innovation of Indigenous brewers. As far back as Stonehenge, Indigenous brewers may have been recycling their spent grain to feed domesticated animals.[41] We know that contemporary brewers are feeding their livestock spent grain from the production of beer.[42]

Breweries such as DogFish Head and Avery Brewing have explored the Indigenous worlds of brewing in creating their own unique signatures of ancient brews. Throughout this book, we have seen how residues taken from the interior of ancient ceramic vessels from around the world have given researchers a new perspective on the ingredients ancient brewers used to brew their beers.[43] Other archaeological methods examining the pollen, starch, and macrobotanical remains have advanced our knowledge of the pioneering ingredients ancient brewers have used to enhance the flavors, smell, and texture of beers.[44] Contemporary Indigenous brewers offer us a rich foundation of the many ways people brew their beers, strengthening their cultural identity.

Sustainability of Beer Production

Producing beer is not environmentally friendly.[45] The environmental drawbacks of beer production and distribution are high rates of greenhouse gas emissions, water usage, and solid waste and a large carbon footprint.[46] However, the rise of the craft beer industry has made significant progress in reducing the amount of waste, which has led to a "greening" throughout the industry. Sustainable movements by the craft beer industry include reducing water and energy use, increasing energy efficiency, and using organic and/or local ingredients, all of which has lowered their overall carbon footprint.[47] Sustainability has three pillars, sometimes referred to as the "Three Ps," consisting of economic viability (profit), environmental responsibility (planet), and promoting social equity (people).[48]

In 1873, there were a staggering 4131 working breweries in the United States, but then Prohibition crushed the craft beer industry and in 1980 there were only eight breweries in operation.[49] The craft beer industry in the United States skyrocketed to 8386 breweries in 2019.[50] Despite the growth of the craft beer industry and its large carbon footprint and other sustainability challenges facing modern brewers, the craft beer industry has met these challenges with innovative and creative solutions. The majority of craft beer breweries collect their spent grains and donate it to farmers to feed animals.[51] 3 Daughters Brewing in St Petersburg, Florida, sells their spent grains to local cattle ranchers and then donate the money they make from the sale to the St Petersburg Free Clinic, the largest emergency food distributor in Pinellas County.[52] The St Petersburg Free Clinic then distributes the food to 65 partner agencies, from food pantries to childcare programs. A co-owner of 3 Daughters brewery says of the program, "It checks all the boxes: sustainability, giving back to the community, and reducing waste."[53]

Breweries use seven pints of water to produce one pint of beer, and to grow enough barley to produce one liter of beer takes 298 liters (78 gallons) of water.[54] Many of the craft beer breweries have reduced their amount of water to produce beer to one to three pints of water.[55] Drought-prone regions of the American Southwest and West are especially vulnerable to water shortages. During the latest California drought, the city of Chico asked the Sierra Nevada brewery to reduce its water consumption by 30 percent.[56] The brewery negotiated with the city to reduce their landscaping and restaurant water consumption but not for their beer production. Sierra Nevada is one of the more sustainable breweries in the country.[57] It built its own wastewater-treatment plant, which not only reduces the amount of water it uses but has a biogas-recovery system that reduces its natural gas needs.[58] America's oldest brewery, D.G. Yuengling & Sons, began back in 1829 and obtained its water from a spring-fed reservoir in the city of Pottsville, Pennsylvania.[59] In 1860, the brewery purchased the reservoir, which they used for almost another 100 years for their Pottsville brewery. Yuengling's Tampa and Pottsville breweries have built their own wastewater pretreatment plants.[60] Yuengling's Pottsville brewery also harnesses the methane gas from brewing their beer and uses it to generate electricity for their wastewater plant and gift shop.[61] Over the past 10 years, the beer industry has reduced its intake of water by 30 percent, and their sustainability efforts not only benefit the environment but also lower the cost of making beer.[62]

Brewers have reduced their overall energy consumption by recovering heat from the wort cooling or keg water systems.[63] Implementing a series of energy-efficient technologies, from using insulated hot water, steam, and refrigerant pipes, to installing efficient fermentation vessels and storage tanks, to using alternative sources of energy has allowed breweries to significantly reduce their overall carbon footprint. Brewers believe that by implementing sustainable technologies they can increase their profits.[64] Sierra Nevada's brewery produces half of its energy from solar arrays.[65] The Yuengling Breweries capture the carbon dioxide that is produced by the yeast fermentation process.[66]

Brewers have been working to use local ingredients even in places where ingredients have not been able to grow. Brewers throughout the United States have led the way in the use of local ingredients such as hops, grains, and yeast.[67] In Florida or any other subtropical region, hops are not a crop that one would think could grow in such a hot and humid climate with a sandy soil.[68] Since 2015, the University of Florida and a group of independent hops growers have been testing whether hops could grow and survive in Florida. Sinsuke Agehara from the Gulf Coast Research and Education Center and University of Florida Environmental Horticulturalist professor Zhanao Dengs have discovered that hops can survive in the Florida climate.[69] Now the goal is to grow high-quality hops that Florida and other breweries around the country can use to make exceptional beer. Most of the hops grown in the United States comes from the states of Washington, Oregon, and Idaho, where the sun shines 16 hours a day in the summer, but they can grow only one crop per year. In Florida, the summer sun is up for only about 12 hours a day, but because of the warm weather, two hop crops can be grown per year.[70] To compensate for the shorter amount of sunlight, Agehara added LED lights, and in 2020, only the first season, the yield reached 60 percent of a normal Pacific Northwest hop yearly yield. More than 15 Florida breweries have used the Florida hops, and 3 Daughters Brewery in St Petersburg, Florida, has begun making 60 barrels of pale ale and smaller amounts to brew a lager using hops grown at the research center.[71] They brewed a golden pale ale that they named "That's Deng Good" after University of Florida professor Denghas, with the ale exhibiting "a blend of caramels, citrus fruits, and earthy bitterness from the fresh hop in the kettle."[72]

The ubiquitous plastic six-pack ring used for distributing beer has caused serious damage to wildlife and to the environment. Viewing this destruction firsthand led to the innovative invention of the edible six-pack rings by Saltwater Brewery, located in Delray Beach, Florida.[73] They took the spent grain of barley and wheat and created an edible six-pack ring that wildlife can actually digest safely. Not only is the six-pack ring edible, it is biodegradable. This is the first edible and biodegradable packaging in the beer industry.[74] These technological achievements are part of a long continuum of creation that stretches back to 11,000 BCE. Technological achievements such as these create a strong solidarity among local communities and energize individuals to seek more education in order to brew better home beers, obtain employment in a brewery, or start their own brewery.

Craft Beer University

In the past decade, there have not been enough trained people to meet the demands of the growth of the craft beer industry. Breweries, both small and large, have reached out to their local universities and colleges to assist in establishing brewing arts certificates. The University of South Florida St Petersburg (USFSP), for example, established its brewing arts certificate when Mike Harting of 3 Daughters Brewery needed more qualified staff to meet the amount of beer 3 Daughters was seeking to produce. He met with Frank Biafora, who was dean of the College of Arts and Sciences at USFSP, and they agreed to develop a curriculum that included USFSP faculty and Tampa Bay brewers. However, before we could establish this certificate we had to convince the USFSP administrators. I started researching the economic impact of breweries and was delighted that the University of Florida had developed an economic impact document on the growth of craft breweries in Florida.[75] When we met with the university administrators, there was some trepidation about establishing a beer certificate at the university, but once we could document that the amount of revenue brewers generate in employing people from the community and how it enhances the cultural atmosphere of the region, we were able to receive a green light on the certificate. This certificate is now in its seventh year and is recognized by the Masters Brewers Association of America (MBAA), and everyone who has graduated with this certificate has either been employed in a brewery, started their own brewery, or begun to improve the beer that they privately produce. The certificate has now been partnered with 15 breweries. Over the years, the Yuengling brewery in Tampa has established scholarships for veterans, women, and minority brewers. The Dunedin Brewers Guild, Rapp Brewery, and Beer Kulture have also established scholarships.

The first 10 weeks of the certificate course includes a detailed online curriculum of the archaeology and Indigenous knowledge of beer, beer styles, water chemistry, hops, malting and mashing, wort, microbiology and fermentation, packaging, safety, and the business of breweries. Students participate in brewery tours, create their own beer recipes, and then work with a local brewery to receive hands-on training. Some of the recipes have included making beer from stale bread, similar to ancient Mesopotamian beer, as well as using chocolate from the Pinellas Chocolate Company, a local Tampa Bay company (see recipes later in this chapter). Future beer recipes may include fruits from local farms and reclaimed water. We have had several international students who have partnered with one of their local breweries so that the online brewing curriculum can be taught to students in any area of the world where there are craft breweries. This curriculum has also begun offering advanced brewing techniques courses, including: starting a brewery quality program on a budget; sour beer and barrel- aging; cider and mead making; yeast wrangling from the craft brewer; and data analysis for the craft brewer. Eleven other brewing certificates and degrees are recognized by the Masters Brewers Association of America, ranging from four- and two-year BA and associate programs to certificates.[76] The growing relationship between breweries and degree programs, with the interest of researchers in interpreting ancient brews, has created innovative and delicious beers. In the following, we offer six successful beer recipes that have a cultural connection to either ancient or contemporary Indigenous beers.

Past to Present Recipes

These beer recipes were created by breweries or university professors working with students and archaeologists to re-create beer based on their own research.

Dr. Jiajing Wang from Stanford University, who specializes in the origins of agriculture, created the first two beer recipes. The first recipe is based on ancient East Asian beer, and the second recipe is adapted from the contemporary Indigenous society of Jívaro, who live in the Amazon and brew a beer from the cassava root.[77] The end result for the barley and wheat beers was a fruity and a bit sour taste, whereas the maize chicha beer had a sweet taste and aroma. The manioc chicha exhibited a sour and fruity taste.[78]

Barley/Wheat/Rye/Millet Beer

Dr. Jiaging Wang

Material needed:

Up to 5 pounds of cereal grains (use one type of grain, either barley, wheat, rye, or millet)
A large container (see right) with flat surface (for malting)
A clean vessel with lid for mashing and fermentation

Roller/pestle for milling/crushing grains
Low-temperature oven
Water

Method and sampling procedure:

1. Malting:
    1) [Steeping] Soak the grains in water for 24 hours.
    2) [Germination] Drain and spread the grains out on a clean surface for germination; place the grains in a cool place (the ideal temperature is 64°F, or about 18°C) and without much light.
    3) [Germination] The grains should germinate in 2–5 days. During this time you need to turn and move the grains about every 12 to 24 hours. You should also spray a light mist over them to keep them moist, though not wet. Note that the time for the grains to germinate properly can vary, and some grains may remain ungerminated. Keep the surrounding environment clean to prevent bacteria from settling on the grains.
    4) [Germination] Examine the grains to see when germination is complete. You want to look for the new growth stemming out from the end of the kernels and up the back of the grain. This growth is called the "acrospire." When the acrospire is roughly the same length as the kernel, the malt is fully modified. If you let it grow longer than the kernel size, the malt is "overmodified." If it is shorter than the kernel size, the malt is "undermodified." Neither under- nor overmodified malt is desired.
    5) [Drying] In modern brewing, the germinated grains go through a step, called "kilning," that includes drying and curing. First, the malts are dried at a low heat over a long period of time to drive off the moisture. This is typically done at 90 to 100 degrees F, with constant air movement, for about 2 days. In the curing stage of kilning, the temperature is raised to 172 to 220 degrees F for another day and a half to 2 days. But for the purpose of this experiment, we skip the curing stage.

2. Milling
   1) Mill or crush the malt with a roller or equivalent. Properly crushed malt has the outside malt husk torn off the seed but left somewhat intact; it should not be torn to bits. The barley kernels should be cracked into large pieces and the hulls should remain largely intact.
3. Mashing
   1) Add water to the clean vessel you prepared ("mash tun"). Measure the amount of water, usually between 1.2 and 1.5 qt./lb. of malt. The water should be around 165–167°F (74–75°C).
   2) Slowly stir in the milled malt, being careful to completely stir it in as you add it to the mash tun. When all the malt is stilled in, check the temperature; it should be between 148 and 154°F (64–68°C). If the temperature is too low, add hot or boiling water to bring it up to the proper temperature.
   3) Hold at this temperature for 1 h.
   4) The mashing should be done after an hour. Now your mash consists of two parts. The liquid part is called "wort" (the German word for unfermented beer), and the other is spent grains.
4. Fermentation
   1) Seal or put a lid on the vessel and place it in a cool spot (out of direct sunlight) to allow fermentation to take place.
   2) Fermentation will complete in less than a week.[79]
5. Use a small mouthed pot for beer storage.

Yucca (Manioc) Beer

Dr. Jiajing Wang

Materials needed:

Yucca (Manioc) roots, 1 large or two small, about 1 pound total
Knife or vegetable peeler
Large cooking pot with lid
Kitchen stove
Optional: large pitcher
Water

Method:

Preparing the mash:

1) Peel and wash the yucca tubers. Using a vegetable peeler or sharp knife, remove the rind—peel down to the white part of the tuber.
2) Cut the tubers into chunks or disks. Put the yucca in a pot and cover with about three or four inches of water. Place the pot on the stove.
3) Bring the water with the yucca to a boil. When the yucca is soft (after 30 min. to an hour), remove the pot from the heat and allow it to cool.
4) Remove the yucca from the pot. Don't pour out the water from the pot.
5) Put handfuls of cooked yucca in your mouth and chew it well, spitting it out into the water in the pot. Continue until you've chewed all of the cooked yucca.
6) Top off the pot with more water until it is almost full. Put the pot back on the stove and bring it almost to a boil, then immediately reduce heat to simmer (medium-low) and cook for about an hour, stirring every so often.
7) Remove from the heat and allow to cool. (Tip: you can now transfer the mash to a <u>clean</u> pitcher large enough hold it.)

Fermentation

1) Put the lid on the pot (or cover the pitcher well) and place it in a cool spot (out of direct sunlight) and allow fermentation to take place.
2) Fermentation will be complete in less than a week.[80]

Bread High

Ty Weaver

Ty Weaver from 3 Daughters Brewery and the USFSP Brewing Arts Program created a beer using stale bread similar to what has been described in ancient Mesopotamia.[81] The recipe was picked as the best beer among the USFSP students' entries for this specific cohort of students. The goal was to create a malty brown ale, and it won the competition because it was low in IBUs. The Bread High beer is copper in color and hazy, with an off-white head. Aroma is malty and bready, with no hops detectable. It has a malty flavor and a clean, dry finish. There is an aspect of bread or toast but almost no hop bitterness, just enough to balance the sweetness of the malt. It has a medium body with a smooth texture and medium carbonation.[82].

Batch size: 5.5 gallons

OG: 1.052
FG: 1.013
ABV: 5.25%
IBUs: 20
Boil: 60 minutes

Grain:

Pale 2 row – 8lbs (60%)
Crystal 60L – 8oz (4%)
Honey malt – 8oz (4%)
Chocolate malt – 4oz (2%)
Stale Bread – 3.5lbs – (27%)
Rice Hulls – 6oz – (3%) (Optional if you have lots of time on your hands)

Hops:

Summit – 0.5 oz at 60 minutes
Tettnanger – 2 oz at 10 minutes
Tettnanger – 1 oz at Flameout

Yeast:

Neutral ale yeast – US05, 1056, WLP001

Other:

Whirlfloc – 1 tablet at 5 minutes

Cut the stale bread into small pieces. Mash grains and bread with 4.5 gallons of water at 155°F for 60 minutes. Most of the bread will probably float to the top of the mash initially, but it should look like a normal mash by the time you start to sparge. Sparge with 5 gallons of water at 168°F to collect about 7.5 gallons of wort. Boil for 60 minutes, adding hops and whirlfloc as noted above. Chill the wort to 70°F, pitch the yeast, and ferment until gravity stabilizes around 1.013.

Gallo-Roman Beer

Lisa C. Kahn

Dr Lisa C. Kahn from George Mason University has created a Gallo-Roman beer based on carvings found from the excavated Roman site of Grand in northeast (Grand Est) France. There are no written recipes for beer in the western Roman world, but three carved reliefs were found in the town of Grand. These carved reliefs demonstrate the importance of beer in ancient Gaul.[83] "The brew was pale amber in color, lightly carbonated with a malty aroma and a toasty flavor of apples. When berries or other additives, such as coriander, are included, the flavor can be fruitier or spicier."[84]

Experiments in re-creating the process:

About 5 lbs of 2 row pale malted barley. Barley can be malted by allowing it to germinate and then drying. Lightly toasting it gives it a richer flavor. Slow roasting in a pan will achieve this.

Mill the barley to moderately fine to coarse grind.
    The ancients may have made a dough with the grain and a small amount of water and then added the "foam" from previous batches of beer. This foam we know is the yeast. Allow the dough to rise for 8–12 hours and then form the cakes. Lightly bake the cakes, slowly heating them for about 10 hours and not exceeding 140 degrees F (70 degrees C), since higher temperatures will kill the yeast. They will be the yeast cakes needed for subsequent batches of beer. For an initial batch,

simply wort the beer as described below, and when the wort is finished and cooled, pitch in some yeast.

The worting process: Heat the milled grain in about 2 gallons of water. Slowly, for 1–2 hours, bring the water to no higher than 170 degrees F. (76 degrees C). Hold it at this temperature for an additional hour.

Sparge: use about half a gallon of hot water.

Sift: sift out the spent grain.

Fermentation: Place the wort in a closed container; the ancients used ceramic or metal jars. In several days, when fermentation has ceased, strain off the yeast and pour the wort into another closed jar or barrel for secondary fermentation. Experiment with adding fruits, such as raspberries, to the fermentation. Allow at least two weeks for the secondary fermentation.

Note: Hops were known in some areas of the ancient world, but it's not known if they were added in brewing in northern Gaul.

The resulting beer should be slightly carbonated and tasty[85]!

Mead/Metheglin

Kimberley G. Connor

Kimberley G. Connor from Stanford University adapted the recipe from Sir Kenelme Digby's 1669 mead/metheglin recipe. Digby was a seventeenth-century diplomat and scientist who traveled throughout Europe collecting fermentation recipes. His secretary published 120 different recipes posthumously, the majority of which are for meads. This recipe is reduced down by a quarter, since it was originally created for Digby's extended family, servants, and retainers. This recipe uses the Old English Ale Gallon measurement of 4.621 L, an Avoirdupois pound of 454 g, and a liquid pint of 578 ml.[86] The mead exhibited a honey-gold, light and refreshing, quite sweet with a hint of spice from the ginger.[87].

Digby's original 1669 recipe states:

Take four gallons of water and set it over the fire. Put into it, when it is warm, eight pounds of honey; as the scum riseth, take it clean off. When it is clear, put into it three nutmegs, quartered; three or four races of ginger, sliced; Then let it boil a whole hour Then take it off the fire and put to it two handfuls of ground malt; stir it about with a round stick, till it be as cold as wort, when you put yest to it. Then strain it out into a pot or Tub, that hath a spiggot and faucet, and put to it a pint of very good Ale-yest; so let it work for two days; Then cover it close for about four or five days, and so draw it out into bottles. It will be ready to drink within three weeks.[88]

1.02 gal water
2 lb honey
¾ of a whole nutmeg, cut into 3
3 oz ginger, sliced
0.71 oz 2 row barley, malted
1 packet Lalvin D-47 wine yeast, dissolved in ½ cup of warm water with 1 tsp sugar

1. Sanitize all equipment and surfaces before starting. Heat the water to 140°F in a large stockpot, then add the honey and stir to dissolve. Bring to a simmer, skimming off any scum, which rises to the top. When no more scum rises, add the nutmeg and ginger and simmer for an hour.

2. Crush the malted barley gently using a rolling pin, then add it to the water and honey mixture. Remove the stockpot from the heat, and allow to cool to 98°F. You can speed this process up by placing the stockpot in a tub of cool water.

3. When the liquid reaches 98°F, strain the liquid using a sieve and transfer the liquid to a sanitized 2-gallon fermenter. Stir in the rehydrated yeast, close the fermenter, and add an airlock. Ferment for 7 days.

4. After 7 days of fermentation, bottle in sanitized plastic beer bottles. Leave in a cool, dark place for several days

until the bottles feel firm, then refrigerate them (see notes). Mature for up to 21 days total. Serve chilled.

Notes:

This recipe is for a short mead, and it is not fermented out, which means that fermentation will continue in the bottles. If the pressure builds up too much, the bottles will explode. Using plastic bottles means that if this happens you won't have broken glass everywhere. Monitor the bottles closely by squeezing them to feel how firm they are. Once there is little to no give in the bottles, you can slow the fermentation process by refrigerating them. There is still a risk of explosion, so drink them within a few weeks![89]

Astral Revelations

Vinny Giusto, Jake Berman, and Joshua Garman

Joshua Garman and Vinny Giusto from Hidden Springs Ale Works in Tampa and the USFSP Brewing Arts Program in conjunction with USFSP Brewing Arts student Jake Berman brewed up an imperial oatmeal stout using Pinellas Chocolate Company cocoa nibs and locally roasted Ethiopian Yirgacheffe coffee. The intensity of the beer was balanced by the addition of lactose to the boil as well as the vanilla added to the coffee recirculation. They also used a "clean fermenting yeast" of London 3, allowing "the dark chocolate and coffee flavors to shine and give the beer a silky texture with a medium-full body."[90] Vinny Giusto, Hidden Springs head brewer, recommends using high-quality chocolate and coffee.

Grain bill: 14 lbs maris otter, 3 lbs and 8 oz flaked oats, 1 lb 4 oz of chocolate malt, 1 lb crystal 60 malt, 1 lb carafa 3 special type malt, and 4 oz roasted barley.

Hop bill: 1 oz Nugget hops and 1 oz Willamette hops.

Adjuncts: 1 lb lactose, 6 oz fresh, locally roasted, whole bean coffee and 8 oz fresh locally roasted cacao nibs.

Process: Mash grains in 5 gal filtered or spring water at 153 degrees F for 60 minutes. Heat 3 gallons of filtered or spring water to 175 degrees F for the sparge. Lauter into your boil kettle and sparge until you have collected 7.5 gallons of wort.

Boil your wort vigorously for 2 hours or until your OG reads 1.112 points gravity, add 1 oz Nugget hops at 60 minutes until flameout, add 1 oz of Williamette hops at 20 minutes until flameout, and finally add 1 lb lactose at 10 minutes until flameout.

Cool your wort until it has reached 67 degrees F, and knock out into a pre-sanitized fermenter with 1 Pure Pitch pack of White Labs Irish Ale yeast (WLP 004). Aerate your wort either with an O2 stone or by shaking and rolling your fermenter vigorously. Ferment your wort at 67 degrees for 12 days or until it is completely fermented out, which should be 1.038 points standard gravity (about 9.7% ABV).

If your beer is above or below that mark, you need not worry too much as there is variance in the viability of your yeast, as well as other variables of your fermentation. Most importantly, the beer should taste good to you. At this point, add 6 oz freshly and locally roasted whole bean coffee and 8 oz freshly and locally roasted cacao nibs of your choosing to a sanitized secondary vessel and purge with CO2.

Then rack your beer into this vessel and allow to rest for 1–2 days, but no longer. You can then bottle or keg your beer and carbonate to 2.3 volumes of CO_2 or to your liking.[91]

Future Ferments

This book examined the ancient and contemporary Indigenous beer, and it is the knowledge and skill of modern Indigenous brewers that offer archaeologists a diverse perspective on how beer impacts people's lives in the past and present.[92] I began this book by describing why I began to study Indigenous beers, and it was the ethnographic present living with the Gamo in southwestern Ethiopia that opened my mind to the essential impact beer has had on the ancient and modern worlds.[93] The sophistication and variety of beers from Indigenous brewers is now inspiring excitement in the craft beer industry. Craft brewers are imparting new recipes spawned from the achievements of generations of Indigenous brewers. The decades of mass-produced, unfulfilling beer is now coming full circle, and we have our ancestors to thank for passing down these flavorful and interesting brews.

Acknowledgments

The idea of this book began a long time ago, and many people have inspired me along the way. The Gamo people, who allowed me into their communities to wander around and learn about their lives and listen to their stories, gave me the knowledge about the importance of Indigenous beer. I will be eternally grateful for their hospitality and openness. It is nice to have a second home, and Ethiopia is this place for me. I thank each and every one who has helped me along the way while I have conducted my research there.

I want to thank all of the researchers (Matthew D. Adams, Masahiro Baba, Luis Jaime Castillo, Lindy Crewe, Ayman Damarany, Oliver Dietrich, David Edwards, Renee Friedman, Augustin F. C. Holl, Li Liu, Desmond Morris, Dani Nadal, Frederic Nitschke, Vecihi Özkaya, Oscar Gabriel Prieto, Ryan Williams) who collaborated and shared their photos for this book. Their generosity tells me that the archaeologists and anthropologists who work on beer understand the meaning of community, and I hope to be able to reciprocate one day.

I profoundly appreciate and thank the brewers, Jake Berman, Kimberly G. Connor, Joshua Garman, Vinny Giusto, Lisa C. Kahn, Jiajing Wang, and Ty Weaver, who provided recipes based on ancient ingredients. They were extremely patient with all my email questions and happily responded.

The USF St Petersburg Brewing Arts Certification Program, led first by Frank Biafora and currently by Jennifer Sedillo, has given me the opportunity twice a year to teach about the ancient and contemporary Indigenous beers to students who are eager to learn everything they can about beer. The lectures gave me the foundation for how to organize this book, and so I thank all the brewers, researchers, and administrators who have supported the USF St. Petersburg Brewing Arts Certificate Program.

I also thank all the people I have met along the way, as I was thinking about writing this book, who gave me positive feedback that they would be interested in reading such a book. They do not know how much those conversations motivated me each day.

The University of South Florida St Petersburg generously supported the research for this book by providing a grant to assist in the book's production and giving me sabbatical time to research and write. A special thanks goes out to Cynthia Brown, the USF St. Petersburg Nelson Poynter Memorial Library Specialist, who helped me immensely by obtaining interlibrary loan books. This was not an easy task during the height of the COVID-19 pandemic when it was difficult to receive books from libraries, but her persistence is very much appreciated.

I want to sincerely thank the two reviewers, one anonymous and Dr. Donna Nash, who unmasked herself so that she could assist me further with the revisions, especially the Meso- and South American chapter. I also thank Kathy Weedman Arthur for reading a draft and listening to me mull over ideas as I was researching and writing the book. She is my muse. They all took the time to carefully read drafts of the manuscript and provide indispensable comments and advice. I take full credit for any inaccuracies that occur in the book.

I want to thank my editor, Stefan Vranka, whose patience with me is deeply appreciated, for having the courage in my vision of the book, and to see it through to the end. Everyone at Oxford University Press has been wonderful to work with, and I sincerely thank Rachel Perkins for producing the book's beautiful cover jacket, as well as Brent Matheny and Gabiel Kachuck for their outstanding editorial and marketing work.

I am grateful for and appreciative of the kind praise from Charles Bamforth, Patrick E. McGovern, and Edward Slingerland. Cheers to each of you!

My field research experience is always enhanced because I am lucky to have Kathryn Weedman Arthur and Matthew C. Curtis as colleagues and friends. My research in Ethiopia has been supported by National Science Foundation Funding, National Endowment for the Humanities, and by the University of South Florida St. Petersburg and Tampa. Permission for my Ethiopian research was given by the Authority for Research and Conservation of Cultural Heritage, which is part of Ethiopia's Ministry of Culture.

My family is always there for me, and for this I am forever grateful. My brother, Jeffrey, and sister, Ellen, listen to me as I ramble on the phone about this or that with my research and writing, and they are always listening and giving me support and love.

I owe my biggest acknowledgment and thanks to Kathy and Hannah, who inspire me each day, make the next day better than the one before, and confirm to me that love is more than a dream.

Appendix: Ancient and Contemporary Beer Ingredient Tables

Table A.1 Asian archaeological sites and regions, time periods, and ingredients brewers have used for beer production.

Place	Time	Ingredients
Raqefet Cave, Israel	11,000 BCE	Wild Wheat or Barley[1]
Jiahu, China	7000 BCE	Rice, Wild Grape, Hawthorn Fruit, and Honey[2]
Lingkou and Guantaoyuan, China	7000–5000 BCE	Broomcorn Millet, Triticeae Grasses, Job's Tears, Snake Gourd Root, Ginger[3]
Mijiaya, China	3000 bce	Broomcorn Millet, Barley from Calcium Oxalate, Job's Tears, Tubers (Snake Gourd Root and Ginger), Lily[4]
Godin Tepe, Iran	3400 BCE	Barley from Calcium Oxalate[5]
Mesopotamia	c. 3000 BCE	Cinnamon[6]
India	c. 800 BCE	Rice, Millet or Barley and Jaggery (Crude Palm Molasses), Blossoms of the Mahuwa Tree[7]
Gordion, Turkey	700 BCE	Barley Beer, Grape Wine, Honey Mead[8]
India	Ethnographic	Millet or Rice[9]

Table A.2 African archaeological sites and regions, time periods, and ingredients brewers have used for beer production.

Place	Time	Ingredients
Deir-el-Medina Amarna	1550–1307 BCE 1350 BCE	Emmer Wheat, Barley, Sycamore Figs[10]
Hierakonpolis	2875 BCE	Emmer Wheat, Lolium sp. and Digitaria sp.[11]
Abydos, El Amarna, and Deir-el-Medina, Egypt	c. 1550-1070 BCE	Barley, Emmer Wheat, Dates[12]
Gamo, Ethiopia	Ethnographic	Barley, Wheat, Sorghum, Finger Millet, Maize, Ginger, Garlic, Pepper[13]
Maane, Burkino Faso	Ethnographic	Red Sorghum[14]
Haya, Tanzania	Ethnographic	Bananas, Sorghum[15]

Table A.3 European archaeological sites and regions, time periods, and ingredients brewers have used for beer production.

Place	Time	Ingredients
Balfarg/Balbirnie, Tayside	Neolithic and Bronze Age site	Meadowsweet Henbane Deadly Nightshade Cabbage Mustard[16]
Barnhouse, Orkney	Neolithic	Barley Unidentified sugars, possibly Maltose[17]
Forteviot, Scotland	2199–1977 BCE	Meadowsweet[18]
Rhum Island, Scotland	c. 1940 BCE	Cereal Pollen, Ling/Scotch Heather, Royal Fern, Meadowsweet[19]
Nandrup, Denmark	c. 1500–1300 BCE	Lime Tree, Meadowsweet, White Clover, Honey (Mead)[20]
Kostraede, Denmark	c. 1300–1500 BCE	Grape, Honey, Birch Resin, Pine Tree Resin, Juniper Myrtle, Bog Myrtle or Sweet Gale, Rosemary, Possibly Wheat, Rye, or Barley[21]

Table A.3 Continued

Place	Time	Ingredients
Jullinge, Denmark	c. 200 BCE	Barley, Bog Cranberry, Lingonberry, Cowberry, Sweet Gale[22]
Havor, Sweden	c. 900 BCE	Grape, Birch Tree Resin, Possibly Wheat, Rye, or Barley[23]
Belgium	c. 500s BCE	Cherry, Raspberry, Peach, Grape, Black Currant, Plum, and Pineapple[24]
Europe	c. 50 BCE	Meadowsweet, Honey[25]
Europe	c. 1000 CE	Hops[26]
Europe	c. 500 CE	**Beer Flavoring Ingredients:** Bugloss, Wormwood, Wild Ginger, Scotch Heather, Limon, St. Benedict's Thistle, Eyebright, Fennel, Wild Strawberry, Herb Bennet, Peppermint, Horse-Heal, Juniper, Bay Tree, Lavender, Marjoram, Lemon Balm, Pennroyal, Spearmint, Oregano, Hart's-Tongue, Norway Spruce, Anise, Silverweed, Wild Cherry, Blackthorn, Oak, Rosemary, Blackberry, Raspberry, Sage, Elderberry, Wood Sanicle, Hedgenettle, Clove, Teucrium, Breckland Thyme, Gypsyweed, Periwinkle, Ginger[27]
Europe	c. 500 CE	**Ingredients for medicinal beers:** Sweet Flag, Caraway, Chinese Cinnamon, Cinnamon, Cardamom, Woodruff, Liverwort, St. John's Wort, Iris, Marsh Labrador Tea, Baldmoney, Nutmeg, Parsley, Polypody[28]

Table A.4 Meso- and South American archaeological sites and regions, time periods, and ingredients brewers have used for beer production.

Place	Time	Ingredients
Cerro Baúl	1000–550 CE	Molle Berry[29] Quinoa, Amaranth[30]
Tarahumara/Rarámuri, Northern Mexico	Ethnographic	Maize, Anomalus Seed (*basiáhuari*)[31]
		Copalquín (used to sweeten the beer if it is too bitter after fermentation) Randia Bark (*papache*)[32]
		Kasará or Kasaráka[33]
		Agave[34]
		Saw-Tooth Candyleaf (*ronínowa*)[35]
		Red Scale Scaly-Polypody and Rio Grande Scaly Polybody added to the beer to increase fermentation and increase the medical qualities of the beer[36]
Monsefú, Peru	Ethnographic	Hoof of a bull[37]
Cochabamba Valley, Bolivia	Ethnographic	Prickly Pear Cactus—cactus fruit added for color Purple amaranth—added for color[38]
Río Beni, near Rurrenabaque, Bolivia	Ethnographic	*Bi* palm—species of *Mauritia*[39]

Notes

Chapter 1

1. Arthur 2002, 2003, 2014a, 2014b.
2. Arthur 2002, 2003, 2014a, 2014b.
3. The term Indigenous follows the United Nations (2021) definition, "Indigenous peoples are inheritors and practitioners of unique cultures and ways of relating to people and the environment. They have retained social, cultural, economic, and political characteristics that are distinct from those of the dominant societies in which they live." I capitalize Indigenous as a symbol of respect, similar to how we capitalize the word "Western."
4. Based on the present research, Australian and North American Indigenous societies did not rely on beer in their past or contemporary societies.
5. Bamforth 2004; Liu et al. 2018
6. McGovern et al. 2013.
7. McGovern et al. 2005.
8. Wang et al. 2016.
9. Myhre 2007.
10. Valdez 2012.
11. Kennedy 1963.
12. Haggblade and Holzapfel 2004.
13. Arthur 2003; Homan 2004; Li et al. 2018.
14. See Dietler 2006:231–232; R. Haaland 2007.
15. G. Haaland 1998; R. Haaland 2016.
16. Bowser and Patton 2004; Willis 2002.
17. Bauer and Stanish 2001:34; Stanish and Bauer 2004.
18. Mehdawy and Hussein 2010:19.
19. Centeno Martín 2018.
20. Fontein 2016:358–359.
21. Kennedy 1963.
22. Hunter 1979:89–90.
23. Jonathan Sauer, as quoted in Braidwood et al. 1953:516.
24. Paul Magledorf, as quoted in Braidwood et al. 1953:519.

25. Helbaek 1953:518; Sewell 2014:3.
26. Yates et al. 2021.
27. Liu et al. 2018.
28. Liu et al. 2018.
29. Chi and Hung 2013; Zhang and Wanli 2010; McGovern 2009:28–42.
30. Edwards 2003; Haaland 1987, 1995, 1999.
31. Smalley and Blake 2003.
32. Arthur 2002, 2003, 2014a, 2014b; McGovern 2009.
33. McGovern et al. 1999; Lagerås 2000.
34. Jonathan J. Powell 2009:xxv.
35. Feeney 1997, as quoted in Bamforth 2002:228.
36. WHO 2018.
37. Vallee 1998.
38. Grigg 2004.
39. Baron 1972:3–4, 6.
40. Baron 1972:8–9.
41. Armelagos et al. 2000.
42. Dietler 2006; Gardiner 1836:266; Green 1999:414; Haggblade 1992:395; Karp 1980:85; Moore and Vaughan 1994:192; Richards 1939:80; Saul 1981:746–747.
43. Bamforth 2002; Dineley 2004:2; Esterick and Greer 1985; Kondo 2004; Steinkraus 1994:261.
44. Haggblade and Holzapfel 2004:339.
45. Meyer 1993:82, as cited in Willis 2002:65.
46. Meyer 1993:82, as cited in Willis 2002:65. Missionaries who worked in the Nyakyusa area of Tanzania from the 1890s to 1916
47. Willis 2002:61. Tanzanian informant.
48. Dietler 2001.
49. 1963.
50. Willis 2002:73.
51. King et al. 2009; Macdonald and Molamu 1999; Ohenjo et al. 2006.
52. McAllister 1993.
53. Madsen and Madsen 1979:43.
54. Arthur 2014a.
55. Arthur 2002, 2003, 2014a; Carlson 1990; Colson and Scudder 1988: Dietler 2006:241; Green 1999; Luning 2002; Wilson 1954.
56. Armstrong 1993:28–42; Joffe 1998:308.
57. Pirazzoli-t'Serstevens 1991.
58. Edwards 2014.
59. Carlson 1990.

60. Dietler 200170–71, 2006:236; Suggs 1996; Haaland 2012.
61. Bowser and Patton 2004.2
62. Allen 2009.
63. Dietler 2001.
64. Allen 2009:38.
65. de Garine 1996:210.
66. Sangree 1962:11.
67. Robbins 1979:371.
68. Karp 1980; Sangree 1962.
69. Carlson 1990.
70. Netting 1964:377, describing the Kofyar people, who live in northern Nigeria.
71. See Dietler 1990:365–370; 2006.
72. Kemp 1986, 1989; Westbrook 1995:157.
73. Hastorf and Johannessen 1993:131.
74. Morris 1979: 28–32.
75. Arthur 2014b.
76. Netting 1964.
77. McDougall 2009; Netting 1964; Lieberman et al. 2020.

Chapter 2

1. Patterson and Hoalst-Pullen 2014:1.
2. See Palmer and Kaminski 2013.
3. See White and Zainasheff 2010,
4. Patterson and Hoalst-Pullen 2014:1.
5. Eaton 2018:54–56.
6. Eaton 2018:56–60.
7. See Hayashida 2008.
8. Arthur 2002; Hayashida 2008.
9. See McGovern et al. 1996.
10. Hayashida 2008:161.
11. Arthur 2002, 2003, 2014a; Dineley 2004; Hayashida 2008.
12. Liu et al. 2018.
13. 1953.
14. Damerow 2012:2.
15. Poelmans and Swinnen 2011:6.
16. Homan 2004.

17. Paulette 2020.
18. Rabin and Forget 1998; Unger 2001.
19. Unger 2001:160.
20. Unger 2001:160; Bennett 1991.
21. Bennett 1991:168.
22. Bennett 1991:168–169.
23. Sozen and O'Neill 2017.
24. Sheinbaum 2018.
25. Dietler 2006:236.
26. Willis 2002:62–63, 67.
27. Liu et al. 2018.
28. Liu et al. 2019.
29. See Liu et al. 2019; Jin et al. 2017; Huang 2000; Jennings et al. 2005:281–282.
30. Liu et al. 2019.
31. Wang et al. 2016.
32. Wang et al. 2016:6444.
33. McGovern et al. 1996.
34. 1993.
35. McGovern 2009:67–68.
36. De Keersmaecker 1996.
37. Damerow 2012:4.
38. Nissen et al. 1993; Damerow 2012:4; Paulette 2020.
39. Paulette 2020:68.
40. Damerow 2012:5; see also Powell 1994:97.
41. Damerow 2012:5; Powell 1994.
42. Damerow 2012:6–8.
43. Damerow 2012:3.
44. Civil 1964.
45. Otto 2014.
46. Zarnkow et al. 2011:49.
47. Woolley 1934, pl. 200, no. 102; Damerow 2012.
48. Homan 2004.
49. Hartman and Oppenheim 1950:1–7.
50. Nissen et al. 1993:36; Pollack 1999:110.
51. Damerow 2012:15; Hartman and Oppenheim 1950:15.
52. Hartman and Oppenheim 1950:15.
53. Hartman and Oppenheim 1950:16.
54. Hartman and Oppenheim 1950:16.
55. Damerow 2012:15.
56. Damerow 2012:16.

57. Paulette 2020.
58. Phillips 2014.
59. Munro 1996.
60. Ghosh et al. 2016.
61. Munro 1996:69–71.
62. See Roy 1978.
63. Roy 1978:311.
64. Takamiya 2008; Wang et al. 2021.
65. Wang et al. 2021.
66. Wang et al. 2021.
67. Attia et al. 2018:78, Fig. 2b; Friedman 2011; Stevenson 2016; Takamiya 2008; Wang et al. 2021.
68. Baba and Friedman 2016:193; Wang et al. 2021.
69. Attia et al. 2018; Friedman 2009:34; Wang et al. 2021.
70. Attia et al. 2018; Kubiak-Martens and Langer 2008; Wang et al. 2021.
71. Attia et al. 2018; also see Stika 1996; Valamoti 2018; Wang et al. 2021.
72. Wang et al. 2021.
73. Samuel 2000:556.
74. see Djien 1982:31; Platt 1964:71; Wood 1994:271.
75. Samuel 2000:556.
76. Arthur 2002, 2003; Odunfa 1985:171; Wood and Hodge 1985:287.
77. De Keersmaecker 1996; Samuel 2000.
78. Samuel 2000:556.
79. Samuel 2000:556.
80. Samuel 2000.
81. Samuel 1996b; Shortland and Tite 2000; Toivari-Viitala 2011.
82. Samuel 1996a.
83. Samuel 1996a.
84. Darby et al. 1977:547; Lucas and Harris 1962:15; Samuel 2000:549.
85. Nims 1958; Spalinger 1988.
86. Samuel 2000:549.
87. Samuel 1996a.
88. 1996a.
89. Samuel 2000.
90. Samuel 1996b.
91. Samuel 2000.
92. Samuel 1996b:9.
93. Samuel 1996b:10.
94. Samuel 2000:552; Palmer 1989:129.
95. 2000:552.

96. Hough 1991:21; Samuel 2000:552.

97. Samuel 2000.

98. Samuel 2000:552–553.

99. Epron and Daumas 1939: pl. 70, top right.

100. Garstang 1907:73, 76, Figs. 61, 62; Samuel 2000:553.

101. Samuel 2000:553.

102. Samuel 2000:553.

103. Samuel 2000:553–554.

104. Samuel 2000:554.

105. Samuel 1996, 2000; Wang et al. 2021.

106. e.g., Kemp 1989:124; Peet 1931:155–156; Samuel 2000:555.

107. Helck 1975, 1977; Samuel 2000:548.

108. Samuel 2000.

109. Cornelius 1989.

110. Daines 2008.

111. Budge 2012:24–29.

112. Marks 2018.

113. Attia et al. 2018; Samuel 2000.

114. 1996a, 1996b, 2000.

115. Marks 2018.

116. Arthur 2002, 2003, 2014a, 2014b; Carlson 1990; Hunter 1979; Luning 2002; Aka et al. 2014.

117. Aka et al. 2014.

118. Reusch 1998:24.

119. Donham 1999:153–155.

120. Hunter 1979:104.

121. Arthur 2002, 2003.

122. Arthur 2014b.

123. Luning 2002.

124. Luning 2002:238.

125. Luning 2002:238.

126. Cilurzo 2012:744; Luning 2002:237.

127. Luning 2002:236–237.

128. Luning 2002:237.

129. Carlson 1990.

130. Carlson 1990.

131. Nelson 2014:9.

132. Koch 2003; McGovern 2013.

133. Nelson 2005:65–66.

134. Nelson 2005:56–57, 60–63; Nelson 2014:11.

135. Nelson 2014:11.

136. Nelson 2014:11–12.

137. Dineley 2004:3.

138. Dineley 2004:4; Stika 1996:66–67.

139. Behre 1999; Nelson 2014:12.

140. Behre 1999.

141. Cool 2006:141–142.

142. See Behre 1999.

143. Behre 1999:36.

144. Behre 1999:37, 39.

145. Behre 1999:42.

146. Behre 1999:42.

147. Behre 1999:43.

148. Unger 2001:110.

149. Behre 1999:39.

150. See Behre 1999:39.

151. Behre 1999:41.

152. Brombacher et al. 1997.

153. Hingh and Bakels 1996.

154. Behre 1999:38.

155. Behre 1999:42.

156. Behre 1999:40; Wilson 1975.

157. Behre 1999:42.

158. Unger 2001:111.

159. Behre 1999:43.

160. Behre 1999:43.

161. Narziss 1984.

162. Behre 1999:43.

163. Behre 1999:43.

164. Behre 1999:43; Pollington 2000:200; Nelson 2005:88; 2014:16.

165. Nelson 2014:17.

166. Nelson 2014:17–18.

167. Koch 2003:135–137; McGovern 2009:147, 152; Nelson 2014:18.

168. Nelson 2014:18; Owen 1841:532.

169. Nelson 2014:18.

170. Nelson 2014:18.

171. Unger 2001:112.

172. Unger 2001:111.

173. Unger 2001:111–112.

174. Unger 2001:111.

206 NOTES

175. Unger 2001:112.
176. Unger 2001:113.
177. Unger 2001:113.
178. Unger 2001:114.
179. Unger 2001:115.
180. Unger 2001:108.
181. Unger 2001:109–110.
182. Unger 2001:110.
183. Unger 2001:109.
184. Cool 2006:143; Unger 2001:109.
185. Unger 2001:110.
186. See Unger 2001:110.
187. Jennings e5t al. 2005; Smalley and Blake 2003.
188. Iltis 2000:36; Smalley and Blake 2003.
189. Mangelsdorf et al. 1967:182.
190. Piperno et al. 2009.
191. 1970:249.
192. Cutler and Cardenas 1947:41.
193. Lumholtz 1898:11.
194. Jennings and Chatfield 2009:200.
195. D'Altroy and Hastorf 1992:265; Goldstein 2005:207; Jennings and Chatfield 2009:218; Morris 1979:28–32; Moseley et al. 2005:17,267; Shimada 1994:144, 169, 208.
196. Moseley et al. 2005:17,267; Williams et al. 2019.
197. Costin 2001:235; Morris and Thompson 1985:90.
198. Anderson 2009:167.
199. Goldstein 2003, 2005; Janusek 2003; Anderson 2009:171–172.
200. Moseley et al. 2005:17,268.
201. Allen 2002; Hastorf 1991; Jennings and Chatfield 2009; Morris 1979:28; Perlov 2009; Orlove and Schmidt 1995:276.
202. Hayashida 2009; Gero 1990, 1992; Moore 1989; Jennings and Chatfield 2009:207–208.
203. Allen 2002:151–153; Jennings and Chatfield 2009:207; Perlov 2009.
204. Jennings and Chatfield 2009:207, 212.
205. Cutler and Cardenas 1947:52.
206. See Hayashida 2008.
207. See Nicholson 1960.
208. González Holguín 1989:18,248, as described by Hayashida 2009:236.
209. Schaedel 1988, as described by Hayashida 2009:236.
210. Jennings and Chatfield 2009:207; Nicholson 1960.

211. Cutler and Cardenas 1947:41.
212. González Holguín 1989:18, 248, as described by Hayashida 2009:236.
213. Cutler and Cardenas 1947:45.
214. Nicholson 1960.
215. Nicholson 1960.
216. Cutler and Cardenas 1947:45.
217. Cutler and Cardenas 1947:45; Jennings and Chatfield 2009:212.
218. Nicholson 1960.
219. Cutler and Cardenas 1947:45; Nicholson 1960.
220. Hayashida 2009:250.
221. Cutler and Cardenas 1947:47.
222. Cutler and Cardenas 1947:51.
223. Nicholson 1960.
224. Cutler and Cardenas 1947:57.
225. Lothrop 1926; La Barre 1938.
226. Lumholtz 1898:11.
227. Farabee 1908:28; La Barre 1938.
228. Cutler and Cardenas 1947:58-5.
229. See Valdez 2012.
230. Uzendoski 2004, 2005.
231. Irigoyen-Rascón 2015:186, 204, 222, 232, 251, 333; Kennedy 1963:622.
232. Kennedy 1963:622.
233. Kennedy 1963:622-633, 1978:113-114.
234. Kennedy 1978:114.
235. See Cunningham 2013; Wylie 1989.

Chapter 3

1. Liu et al. 2018.
2. Smith and Lee 2008.
3. Bertman 2003; Pollack 1999.
4. Centeno Martín 2018.
5. Helbaek 1953:518; Sewell 2014:3.
6. Liu et al. 2018.
7. Yeshurun et al. 2013.
8. Liu et al. 2018.
9. Liu et al. 2018:784.
10. 2018.

11. Liu et al. 2018:787.
12. Hillman 2000; Liu et al. 2018:788.
13. Liu et al. 2018:791.
14. Liu et al. 2018:791–792.
15. Arranz-Otaegui et al. 2018.
16. 2013; also see Wright 1994:243.
17. Hayden et al. 2013:133.
18. Perrot 1966:476.
19. Moore et al. 2000:174.
20. Stordeur and Abbés 2002:583–584, Fig. 12/1–4.
21. Yartah 2004:155, Fig. 18/2, 4–5.
22. Mazurowski 2003:369, Fig. 11/1–2.
23. Özdoğan 1999:59; Dietrich et al. 2019; Dietrich et al. 2020.
24. Stordeur and Abbés 2002:583–584, Fig. 12/1–4.
25. Ozkaya and San 2007: Fig. 6, 15–18.
26. Rosenberg and Redding 2000:50, Fig. 5.
27. Hayden et al. 2013:113.
28. Hayden et al. 2013:133.
29. See Hayden et al. 2013.
30. Ozkaya and San 2007: Fig. 6, 15–18.
31. Özdoğan 1999:59.
32. Rosenberg and Redding 2000:50, Fig. 5.
33. Stordeur and Abbés 2002:583–584, Fig. 12/1–4.
34. Özkaya and Coskun 2009.
35. Benz 2010.
36. Aurench and Kozlowski 2001.
37. Curry 2008; Harlan and Zohary 1966; Nesbitt and Samuel 1996; Heun et al. 1997, 2008; Lev-Yadun et al. 2000; Özkan et al. 2002, 2011; Luo et al. 2007.
38. Dietrich et al. 2012.
39. Dietrich et al. 2012.
40. Dietrich et al. 2012:687.
41. Dietrich et al. 2012.
42. Curry 2008.
43. Banning 2011; Dietrich et al. 2012.
44. Dietrich 2012:681.
45. Dietrich 2012.
46. Klaus Schmidt 2008 in Curry 2008:2.
47. Rosenberg and Redding 2000:57–58.
48. Rosenberg and Redding 2000:50.

49. Rosenberg and Redding 2000:52.
50. Hayden et al. 2013:103; Haaland 2007:172; Willcox 2002.
51. Willcox and Stordeur 2012.
52. Willcox and Stordeur 2012:108.
53. Speth 2015.
54. Atalay and Hastorf 2006; Speth 2015.
55. Atalay and Hastorf 2006.
56. Atalay and Hastorf 2006:309.
57. Hornsey 1999:38.
58. Katz and Voigt 1986:32.
59. Hornsey 1999:31–32; Briggs et al. 2004:190–194.
60. Banning 1998:225, 2011:639; Dineley 2015.
61. Garfinkel 1987.
62. Dineley 2004:23.
63. Rollefson 1983:32.
64. Kenyon 1957; Bar-Yosef and Alon 1988; Simmons et al. 1990.
65. Dietrich et al. 2012.
66. Hayden et al. 2013; Rosenberg and Redding 2000; Stordeur and Abbès 2002:584.
67. Bender 1978; Hayden 1990, 2009.
68. Spielmann 2002.
69. 2009.
70. Dietrich et al. 2012.
71. Benz 2010; Dietler 2006; Ouzman and Wadley 1997; Rosenberg and Redding 2000:44; Spielmann 2002; Wiessner 1996.
72. Spielmann 2002.
73. Ouzman and Wadley 1997; Wadley 1989.
74. McGovern 2009; Chi and Hung 2013.
75. McGovern 2009; Chi and Hung 2013.
76. McGovern et al. 2005.
77. Smith and Lee 2008.
78. Smith and Lee 2008.
79. Smith and Lee 2008:278.
80. Zhang et al. 2004.
81. Matthiessen 2001:165.
82. Chung 2008:116–117.
83. See Chung 2008:117–118.
84. Li et al. 2002.
85. Chang 1980:33–38; Keightley 1978, 1989:171–202, 1994:71–79, 1996.
86. Liu 2004:65.

87. Li et al. 2002.
88. Liu 2004:66.
89. Smith and Lee 2008.
90. Smith and Lee 2008.
91. Smith and Lee 2008.
92. Smith and Lee 2008.
93. McGovern et al. 2005.
94. Liu et al. 2018; Liu et al. 2020; Wang et al. 2016.
95. Liu et al. 2020.
96. See Dietler 2001; Hayden 2001:29–30; Liu 2004:41.
97. 2004:1.
98. Chang 1999:64–65; Postgate et al. 1995:467–468.
99. Linduff et al. 2000.
100. Liu 2000; Underhill 1989, 1994.
101. Liu 1996; Wang et al. 2016; Underhill 2000.
102. Liu 2004:1.
103. Armstrong 1993:56–65; Joffe 1998:308.
104. Joffe 1998.
105. Paulette 2020.
106. Michel et al. 1993.
107. Homan 2004: Sandars 1972:7–8; Kovacs 1989:xxiii.
108. Kovacs 1989:16.
109. Homan 2004.
110. Cooper 2016; Langlois 2016:117–118; De Graef 2018:95; Vance 2020.
111. Vance 2020.
112. Harris 1975; De Graef 2018; Vance 2020.
113. Homan 2004:85.
114. Bertman 2003: 26: Pollack 1999:192.
115. Pollack 1999:188.
116. Pollack 1999:186.
117. Pollack 1999:186; Frankfort 1978; Nissen 1986:191–193; Jacobsen 1987.
118. Pollack 1999:186–187.
119. Michalowski 1994:29.
120. Michalowski 1994:31.
121. Michalowski 1994:33.
122. Michalowski 1994:37.
123. Gelb 1965; Michalowski 1994; Pollack 1999:2.
124. Neumann 1994.
125. Michalowski 1994:28.
126. Pollack 2003:28–29.

127. Pollack 1999:117.

128. Pollack 2003:28.

129. Stol 1994:167.

130. Pollack 1999:120.

131. Pollack 2003:27; Stol 1994:182.

132. Pollack 2003:29–30; Wright and Johnson 1975; Millard 1988.

133. Pollack 2003:32.

134. Bertman 2003:57; Stol 1994:179.

135. Stol 1994:180.

136. Sewell 2014:24.

137. Stol 1994:180.

138. Stol 1994:180.

139. Bertman 2003:28.

140. McGovern 2009:130–135; Miller et al. 2009:916, 921.

141. McGovern et al. 1999.

142. Simpson 2012:154–155; Boer 2018:68.

143. Roller 1983.

144. Ballard 2012; Rose 2012:15.

145. Simpson 2012:149.

146. Rose 2012:13.

147. Sams 1977.

148. Rose 2012:15.

149. Rose 2012:15.

150. Rose 2012:15.

151. Miller et al. 2009; Sams 1977.

152. Haaland 1999, 2016.

153. Sams 1977.

154. McGovern et al. 1999; Rose 2012:15; Simpson 2012:155.

155. Haaland 2011:1.

156. Fuller 2003, 2007, 2009; Haaland 2011.

157. Haaland 2011.

158. Allchin and Allchin 1982:Fig. 10.24 nos. 10, 13, 14; Haaland 2011:5.

159. Haaland 2011:6.

160. Mohan and Sharma 1995:130.

161. McHugh 2014:34.

162. Kenoyer 1998:49, 52–53, 55–62.

163. Kenoyer 1998:58, 64–65, 122–124.

164. Kenoyer 1998.

165. Kenoyer 1998:43, 77.

166. Kenoyer 1998:16, 69, 78.

167. Kenoyer 1998:163; Petrie and Bates 2017.
168. Kenoyer 1998:130, 149–162, 236.
169. Sharma et al. 2010; Singh and Lal 1979.
170. Allchin 1979.
171. Sharma and Mohan 1999:102.
172. Sharma and Mohan 1999:102.
173. Singh and Lal 1979.
174. Singh and Lal 1979.
175. Prakash 1961.
176. McHugh 2014:35.
177. Boesche 2003; McHugh 2014:36.
178. McHugh 2014:36–37.
179. Sharma et al. 2010.
180. McHugh 2014:38.
181. McHugh 2014:38.
182. Sharma et al. 2010:10.
183. McHugh 2014:38.
184. McHugh 2014:38.
185. McHugh 2014:39.
186. Tschurenev and Fischer-Tiné 2014.
187. Phillips 2014.
188. Tshurenev and Fischer-Tiné 2014:4.
189. Phillips 2014.
190. Deuraseh 2003; Phillips 2014.
191. Munro 1996:87–98.
192. Munro 1996:2–3.
193. Takakura 1960.
194. Munro 1996:69.
195. Munro 1938; Seligman 1996:149.
196. Centeno Martín 2018; Munro 1996:10.
197. Munro 1996:123.
198. Centeno Martín 2018:4.
199. Hiwasaki 2000.
200. Munro 1996:87–98.
201. Munro 1996:91–92.
202. Munro 1996:92.
203. Munro 1996:93–94.
204. Munro 1996:97.
205. Hiwasaki 2000.
206. Sharma et al. 2010; Singh and Lal 1979.

207. Allchin 1979.
208. Phillips 2014.

Chapter 4

1. 1953.
2. Marshall and Hildebrand 2002.
3. Huysecom et al. 2009.
4. Dunne et al. 2012, 2016.
5. Dunne et al. 2016; Haaland 2007; Huysecom et al. 2009.
6. Edwards 2003; Haaland 1992, 2006, 2007.
7. G. Haaland 1998; R. Haaland 2007.
8. Arthur 2002, 2003, 2014a.
9. Sangree 1962:11.
10. Robbins 1979:371.
11. Karp 1980; Sangree 1962.
12. Wilson 1954; see Willis 2002:64.
13. Darby et al. 1977:531–532.
14. Taylor 2010:16.
15. Taylor 2010:16.
16. Taylor 2010:16.
17. Taylor 2010:26.
18. Taylor 2010:217.
19. Hassan 1988.
20. Hassan 1988.
21. Samuel 1996b.
22. 1993.
23. Geller 1993.
24. Geller 1993.
25. Wang et al. In press.
26. Hendrickx 2008; Baba and Friedman 2016:Fig. 11; Friedman 2020; Wang et al. In press.
27. Michel et al. 1992; Wang et al. In press.
28. Baba In press; Wang et al. In press.
29. Wang et al. In press.
30. Friedman et al. 2011; Wang et al. In press.
31. Princeton University, Office of Communications 2021.
32. Lerner 2018.

33. Janssen 1975:457, 460.
34. Janssen 1975:346–347.
35. Edgerton 1951; Janssen 1975:465; Wente 1961.
36. Dirar 1993; Pope 2013; Haaland 2007; Edwards 2003; Samuel 1996a, 2000.
37. Barth 1967:152–154.
38. Edwards 2014.
39. LeClant 1981:288.
40. LeClant 1981:290.
41. LeClant 1981:312; Sayce 1911:55; Tylecote 1970.
42. Trigger 1969:23–50.
43. Armelagos 2000.
44. Nelson et al. 2010.
45. Armelagos 2000.
46. Cook et al. 1989.
47. Cook et al. 1989.
48. Darby et al 1977:548.
49. Phillipson 1994.
50. Schmidt and Curtis 2000.
51. Pikirayi 2016.
52. Pikirayi 2016.
53. Pikirayi 2016.
54. Pikirayi 2016:102.
55. Pikirayi 2016:105.
56. Fontein 2006:12–13; Pikirayi 2016.
57. Fontein 2006; Pikirayi 2016.
58. Fontein 2016:326, 329.
59. Carlson 1990: Netting 1964; Robbins 1979; Sangree 1962.
60. Arthur 2002, 2003, 2006, 2009, 2013, 2014a,2014b,2014c, 2019; Arthur et al. 2019.
61. Freeman 1997; Halperin and Olmstead 1976.
62. Halperin and Olmstead 1976.
63. Halperin and Olmstead 1976; Sperber 1975.
64. Sperber 1975:215.
65. see Bray 2003a,2003b, Dietler and Hayden 2001, Hayden 2014.
66. 2001:116.
67. de Garine 1996; Dietler 2001; Donham 1999:153–155; Reusch 1998.
68. Arthur 2003.
69. Arthur 2002, 2003, 2014a,2014b.
70. Oura et al. 1982:113.
71. Whalen and Minnis 2009.

72. Edwards 2014.
73. Schmidt and Curtis 2000.
74. Job n.d.
75. Perry 2011.
76. Arthur 2006:84.
77. Dietler 2001:96–99.
78. Reusch 1998:29–32.
79. Karp 1980:85.
80. de Garine 1996.
81. de Garine 1996.
82. Netting 1964:376.
83. Netting 1964.
84. Netting 1964:376.
85. Netting 1964:377.
86. Carlson 1990.
87. Carlson 1990:303–304.
88. Carlson 1990:303–304.
89. Netting 1964:376–377.
90. Roscoe 1923:130, 133; Lindblom 1941:63–64.
91. Berns 1988:71–72; Croucher 2012:12.
92. Berns 1988:71–72.
93. Luning 2002; McAllister 1993:79; van Dijk 2002.
94. Hammand-Tooke 1962:65.
95. Luning 2002:231.
96. Van Dijk 2002.
97. Luning 2002:241.
98. Luning 2002:243–244.
99. Abbink 2002:169.
100. Dietler 2001; Dietler and Herbich 2001.
101. Donham 1999:155.
102. Watson 1998:148.
103. Watson 2009:62.
104. Crush 1992.
105. Hunter 1979:89–90.
106. Netting 1964.
107. Netting 1964:377.
108. Moore and Vaughan 1994:22, 191.
109. Moore and Vaughan 1994:192.
110. Bryceson 2002:24.
111. Haggblade and Holzapfel 2004:12.

112. Bryceson 2002:25.
113. Bryceson 2002:25; La Hausse 1988.
114. Willis 2002:62–63, 67.
115. 2002:64.
116. Willis 2002:58.
117. Willis 2002:57.
118. Willis 2002.
119. McCall 2002.
120. Saul 1981.
121. McCall 2002:96–97.
122. McCall 2002:97.
123. Feldstein and Poats 1990; McCall 2002.
124. Fischer and Mshana 1994.
125. Geisler et al. 1985; McCall 2002.
126. Fieldstein and Poats 1990; McCall 2002.
127. 1990:370–371.
128. 1926:346.
129. Mair 1934:183.
130. Robbins 1979:371.
131. Platt 1964.
132. Bryceson 2002:7; La Hausse 1988.
133. Hardy and Richet 1933; Platt 1955.
134. Bryceson 2002:7; Haggblade 1992:395.
135. Haggblade 1984; Haggblade and Holzapfel 2004; Saul 1981.
136. Haggblade and Holzapfel 2004; Novellie 1963.
137. Haggblade 1984; Haggblade and Holzapfel 2004:2.
138. Bryant 1970; Haggblade and Holzapfel 2004; Platt 1955; Quin 1959.
139. de Garine 1996:210.
140. Haggblade and Holzapfel 2004; Katz 1979; Zammit 1979.
141. Haggblade and Holzapfel 2004:12; Lyumugabe et al. 2012; Richards 1939; van Heerden 1987.
142. Cernea 2009; Stevenson and Buffavand 2018.
143. Stevenson and Buffavand 2018.
144. Mandelbaum 1965.
145. Arthur 2003.
146. Sutton 1977.
147. See Fontein 2016 and Pikirayi 2016.

Chapter 5

1. Barclay 1983; Dickson 1978; Dineley 2006; Dineley and Dineley 2000; Guerra-Doce 2015; Hey et al. 2016; Mulville et al. 2016:146; Rojo-Guerra et al. 2006; Worley et al. 2019.
2. Tresset and Vigne 2011; Zohary et al. 2012:40, 52.
3. Stewart 1974; Sarpaki 2009.
4. Kroll 1981; Zohary et al. 2012:38, 44, 57.
5. Blasco et al. 1999; Zohary et al. 2012:44.
6. Martin et al. 2008; Jacomet 2007.
7. Murphy 1989.
8. Göransson et al. 1995.
9. Oalde et al. 2018; Parker Pearson 2012; Sherratt 1987.
10. Nelson 2005:65–66, 2014; Olalde et al. 2018.
11. Sherratt 1991, 1995:25; Barclay and Russell-White 1993 as cited in Dineley and Dineley 2000; Dineley 2004:viii.
12. Alm and Elvevåg 2013; Alm 2003:409–411.
13. Dineley and Dineley 2000; Pryor 2009.
14. Parker Pearson 2012:7.
15. Richards 1996.
16. Parker Pearson 2012:345.
17. Craig et al. 2015.
18. Craig et al. 2015.
19. Parker Pearson 2012:59, 79–81.
20. Albarella and Serjeantson 2002; Craig et al. 2015; Parker Pearson 2012:3; Wainwright and Longworth 1971.
21. Craig et al. 2015.
22. Craig et al. 2015:1103.
23. Craig et al. 2015.
24. Viner et al. 2010.
25. Dineley 2006; see Chapter 7 for comparison to contemporary craft brewing.
26. Wright et al. 2014; Worley et al. 2019.
27. Dineley 2004:57–59.
28. Richards 2005:131–133, Fig. 5.6.
29. Richards et al. 2016:19–20.
30. Dineley 2004:59.
31. Parker Pearson 2012:117.
32. 1987.

33. Brophy and Noble 2009:24; Olalde et al. 2018; Rojo-Guerra et al. 2006; Thomsenn 1929; Vander Linden 2006.
34. Oalde et al. 2018.
35. Parker Pearson 2012:345.
36. Mukherjee et al. 2008; Parker Pearson 2012:117.
37. Guerra-Doce, 2006a, 2006b.
38. Brophy and Noble 2009:24; Noble and Brophy 2011; Dineley 2006; Hey et al. 2016; Mulville et al. 2016:146; Worley et al. 2019.
39. Hey et al. 2016:282–283.
40. Dineley 2006; Hey et al. 2016; Mulville et al. 2016:146; Worley et al. 2019.
41. Lightfoot et al. 2009; Hey et al. 2016:51–92.
42. Hey et al. 2016:51–92.
43. Brophy and Noble 2012.
44. Brophy and Noble 2009:24; Noble and Brophy 2011; Guerra-Doce 2015.
45. Barclay 1983; Guerra-Doce 2015.
46. Brophy and Noble 2009.
47. Nelson 2005:11–12; Long et al. 1999.
48. Haggerty 1991:91; Dineley and Dineley 2000:138.
49. Nelson 2005:12.
50. Dickson 1978:111–112; Barclay et al. 1983:179–180; Sherratt 1987:96.
51. Dickson 1978:112.
52. Dineley and Dineley 2000:138; Wickham-Jones 1990.
53. Dineley and Dineley 2000.
54. Rojo-Guerra et al. 2006.
55. See Parker Pearson 1993:214–216.
56. See Tilley 1996:243.
57. Rahmstorf 2008; Rojo-Guerra et al. 2006:258, Fig. 9.
58. Rojo-Guerra et al. 2006.
59. Garrido-Pena and Muñoz-López Astilleros 2000; Guerra 2006, as cited in Rojo-Guerra et al. 2006.
60. Dickson and Dickson 2000:81; Sherratt 1987:396; Nelson 2005:12.
61. Felding 2015.
62. Frei et al. 2015; Kristiansen and Larsson 2005:298–299.
63. Felding 2015.
64. Applequist and Moerman 2011; Hasson 2011.
65. Felding 2015; Frei et al. 2015.
66. Alexandersen et al. 1981, as cited in Felding 2015.
67. Frei et al. 2015.
68. Coleman et al. 2017; Valamoti 2018.
69. Valamoti 2018.

70. See Nelson 2005:25–37; 2014.
71. McGovern 2003; Tzedakis et al. 2008; Garnier and Valamoti 2016; Valamoti 2018.
72. Job n.d.; Coleman et al. 2017:111–114, Figs 15b, 18; Valamoti 2018.
73. Job, n.d.
74. Valamoti 2018.
75. Valamoti 2018.
76. See Dineley 2004:4; Stika 1996:66–67 for evidence of other breweries being burned down during beer production.
77. Valamoti 2018:617.
78. Valamoti 2018:619.
79. Valamoti 2018:620.
80. Valamoti 2018:620.
81. Crewe and Hill 2012.
82. Crewe and Hill 2012.
83. Crewe and Hill 2012:223–224.
84. Crewe and Hill 2012:223.
85. Crewe and Hill 2012.
86. Crewe and Hill 2012:224–233.
87. Arnold 1999.
88. Sitka 2011a.
89. Sitka 2011a, 2011b.
90. Sitka 2011b.
91. 2011a.
92. 2011a.
93. Sitka 2011a.
94. Daniels 2012:83; Seidl 2012:687–688.
95. Larsson et al. 2018:1970.
96. Larsson et al. 2018.
97. Helbaek 1938, as cited in Larsson et al. 2019.
98. Helbaek 1966.
99. Viklund 1989, as cited in Larsson et al. 2019.
100. Nelson 2005:65–66; 2014
101. Nelson 2014.
102. Poelmans and Swinnen 2011.
103. See Nelson 2005:78–114.
104. Nelson 2005:82.
105. Brettell et al. 2012; Hooke 1998; Lee 2007:27; Nelson 2005.
106. Hornsey 2012:675.
107. Protz 2012:28–31.

108. Protz 2012:28, 32.
109. Hornsey 2012:175.
110. Nugent and Williams 2012.
111. 2012.
112. Leahy 2007:54.
113. 2011.
114. 2011.
115. Richards 1987:154; Fig. 8a; Nugent and Williams 2012, Perry 2011.
116. Zori et al. 2013.
117. McGovern 2013.
118. See Robinson 1994.
119. Larsson et al. 2019.
120. Robinson 1994.
121. Zori et al. 2013.
122. Byock 2001:67–68.
123. Perdikaris 1999; Zori et al. 2013.
124. Gelsinger 1981:14; Sveinbjarnardóttir et al. 2007:202–203; Zori et al. 2013.
125. Zori et al. 2013.
126. Dineley and Dineley 2000.
127. Murakami et al. 2006.
128. DeLyser and Kasper 1994.
129. Wilson 1975.
130. Wilson 1975; DeLyser and Kasper 1994.
131. Wilson 1975.
132. DeLyser and Kasper 1994.
133. Arnold 1911, as cited in DeLyser and Kasper 1994.
134. DeLyser and Kasper 1994; Wilson 1975.
135. Wilson 1975; DeLyser and Kasper 1994.
136. 1975.
137. Wilson 1975:644.
138. Wilson 1975.
139. DeLyser and Kasper 1994:168; Wilson 1975:644.
140. Wilson 1975.
141. Wilson 1975; DeLyser and Kasper 1994.
142. Edwardson 1952, as cited in DeLyser and Kasper 1994:167.
143. Edwardson 1952; Neve 1976.
144. Wilson 1975.
145. Wilson 1975.
146. Edwardson 1952.

147. Parsons 1940.
148. Edwardson 1952.
149. Wilson 1975.
150. DeLyser and Kasper 1994.
151. Wilson 1975.
152. Monckton 1966.
153. DeLyser and Kasper 1994.
154. Parsons 1940.
155. Pearce 1976.
156. DeLyser and Kasper 1994.
157. Alm and Elvevåg 2013.
158. Matassian 1989:23.
159. Heusinger 1856:12–14, as cited in Alm and Elvevåg 2013.
160. Alm and Elvevåg 2013.
161. Alm 2003:408, 413.
162. Alm and Elvevåg 2013.
163. Molvik 1992:53–55, as cited in Alm and Elvevåg 2013.
164. Alm 2003.
165. Alm 2003:409.
166. Hagen and Sparboe 1998, as cited in Alm 2003:409.
167. Alm 2003:413.
168. Alm 2003:410.
169. Hagen and Sparboe 1998, as cited in Alm 2003:411.
170. Mittag 2014:67.
171. Narziss 1984:351.
172. Nelson 2014:14–15.
173. Narziss 1984:351.
174. Patterson and Hoalst-Pullen 2014; Yool and Comrie 2014.
175. Mittag 2014:67.
176. Rabin and Forget 1998:12.
177. Mittag 2014:67.
178. Rabin and Forget 1998:165; Mittag 2014:67.
179. Mittag 2014:68.
180. Mittag 2014:67, 70.
181. Mittag 2014:70.
182. Mittag 2014:70.
183. Mittag 2014:70–71.
184. Mittag 2014:71.
185. See De Keersmaeker 1996.
186. Hartman and Oppenheim 1950:15; Mittag 2014:67, 69; Samuel 2000.

187. Mittag 2014:70.
188. Mittag 2014:69.
189. De Keersmaeker 1996; Mittag 2014:70.
190. De Keersmaecker 1996:78.
191. Mittag 2014:70.
192. Mittag 2014:71.
193. Dornbusch and Oliver 2012:660–664; Mittag 2014:71.
194. Mittag 2014:71.
195. Mittag 2014:72.
196. Smith 2013.
197. Smith 2013.
198. Mittag 2014:72.
199. Mittag 2014:72.
200. Pryor 2009.
201. Curtis 1838:30.
202. Curtis 1838:24–25.
203. Pryor 2009.
204. Mittag 2014:67.
205. Mittag 2014:69.
206. Hampton 2012:650; Mittag 2014:68.
207. Pearson 2017.
208. Mittag 2014:68.
209. Tonsmeire 2014:2; Cilurzo 2012:744.
210. Tonsmeire 2014:2; Kelley 1965.
211. Tonsmeire 2014:2.
212. Tonsmeire 2014:3.
213. Tonsmeire 2014:3.
214. Tonsmeire 2014:3–4.
215. Parker Pearson 2012.
216. Dickson 1978; Dineley 2004; Guerra-Duce 2006a; Nelson 2005; Rahmstorf 2008; Sherratt 1987.
217. Coleman et al. 2017; Job n.d.; Valamoti 2018.
218. Dickson and Dickson 2002; Sherratt 1997; Nelson 2005.
219. Perry 2011.
220. Alm 2003.
221. Wilson 1975; DeLyser and Kasper 1994.
222. Mittag 2014.

Chapter 6

1. Piperno et al. 2009; Smalley and Blake 2003.
2. Burger 2008; Moore 1989; Morris 1979; Moseley et al. 2005.
3. Hastorf 2003; Morris 1979; Swenson 2007:65; VanPool and VanPool 2019; Whalen and Minnis 2009; Williams and Nash 2006.
4. Carod-Artal 2015.
5. Logan et al. 2012; Moore 1989; King et al. 2017.
6. Hayashida 2008; Nicholson 1960.
7. Bowser and Patton 2004; Kennedy 1963; Perlov 2009; Uzendoski 2004.
8. Allen 2002; Hayashida 2008; Orlove and Schmidt 1995; Morris 1979.
9. Bray 2009; Burger 2008; Murra 1980; Shimada 2001; Swenson 2006; Williams and Nash 2006.
10. Allen 2002; Bolin 1998; Jennings and Bowser 2009; Kaulicke 2015; Meddens et al. 2010.
11. Morris 1979; Hayashida 2019; Jennings 2005; Zawaski and Malville 2007.
12. Piperno et al. 2009.
13. Hastorf and Johannessen 1993:429; Norr 1995; Smalley and Blake 2003:684.
14. Smalley and Blake 2003:675, 677–678.
15. Iltis 2000:36, as quoted in Crosswhite 1982 and elaborated on by Smalley and Blake 2003.
16. Smalley and Blake 2003:678; Iltis 2000; Piperno and Pearsall 1998:162.
17. Mangelsdorf et al. 1967, as cited in Smalley and Blake 2003:682.
18. Mangelsdorf 1974:156, as cited in Smalley and Blake 2003:682.
19. Smalley and Blake 2003:682.
20. MacNeish et al. 1981:134–135; MacNeish et al. 1983:166–170; Smalley and Blake 2003:682.
21. Piperno 1988; Smalley and Blake 2003:687.
22. Pearsall 2002; Piperno 2003; Staller and Thompson 2002; Staller 2003; Smalley and Blake 2003:687.
23. Smalley and Blake 2003:687.
24. Gamboa 2002:41, as cited in VanPool and VanPool 2019.
25. VanPool and VanPool 2019; Whalen and Minnis 2009.
26. King et al. 2017; VanPool and VanPool 2019.
27. VanPool and VanPool 2019.
28. Grimstead et al. 2013; Rakita and Cruz 2015; VanPool and VanPool 2019.
29. Fish and Fish 1999; Pitezel 2011; Douglas and MacWilliams 2015:146; VanPool et al. 2005; VanPool and VanPool 2019.

30. King et al. 2017; Whalen and Minnis 2009:171–176.
31. Whalen and Minnis 2009.
32. Whalen and Minnis 2009:174–176; Minnis and Whalen 2005:53.
33. Noneman et al. 2017; VanPool et al. 2016, as cited in King et al. 2017.
34. King et al. 2017.
35. Boyadjian et al. 2007:1622.
36. Boyadjian et al. 2007.
37. King et al. 2017.
38. King et al. 2017.
39. Kennedy 1978:114–115.
40. Kennedy 1978.
41. 2009.
42. Kennedy 1963, 1978.
43. Kennedy 1963:623.
44. Kennedy 1963:623.
45. Kennedy 1963.
46. Kennedy 1963:624.
47. Kennedy 1963:626.
48. Burger 2008; Rick 2005.
49. Burger 2008:684.
50. Reinhard 1985; Burger 2008.
51. Rick 2005:72.
52. Burger 2008:685; Rick 2005.
53. Burger 2008:687–688.
54. Burger 2011; Wilke and Wunn 2019:9–10.
55. Burger 2011; Wilke and Wunn 2019:9–10.
56. Rick 2006; Burger 2011; Torres 2008.
57. Burger 1992; Lumbreras and Amat 1965; Tello 1960; Rick 2005:Figs. 1 and 2.
58. Burger 2008:690; Kolar 2017; Moseley 1992:155.
59. Moseley 1992:155–156.
60. Donna Nash, 2021, personal communication; Moseley 1992:155–156.
61. Rick 2005.
62. Burger 2008:688.
63. Burger 1984:172–187; Burger 2008:692; Burger and Van der Merwe 1990: Lumbreras 1993:Figs. 618, 618a, 619.
64. Burger 2008:688; Burger and Van der Merwe 1990:92.
65. See Rick 2005.
66. See Burger 2008.
67. Butters and Uceda 2008:707.

68. Moseley 1992:179–183; Bray 2009; Dillehay and Kolata 2004; Morris 1979.
69. Moseley 1992:166.
70. Shimada 1994:221–224, 2001:187, 192; Swenson 2006,
71. Shimada 1994:221–224, 2001:187, 192.
72. Swenson 2006.
73. Swenson 2006:Fig. 5.4.
74. Swenson 2012b.
75. Swenson 2012a:9
76. Swenson 2012a:10.
77. Swenson 2012a:11–16
78. Swenson 2012a:15.
79. Swenson 2012a:16.
80. Swenson 2012a:18–19
81. Swenson 2012a:179.
82. Swenson 2012a:179: Ramirez 1996, 1998.
83. Swenson 2012a:179–181.
84. Bawden 1996, as cited in Butters and Uceda 2008:723; D'Altroy 2015a:395–401; Dell 2010:1867; Shimada 1994.
85. Moseley et al. 2005; Nash 2012, 2019; Nash and deFrance 2019; Williams and Nash 2006; Williams et al. 2019.
86. Nash 2012; Nash and deFrance 2019.
87. Nash 2012; Nash and deFrance 2019.
88. Nash 2010; Nash and Williams 2009; Nash and deFrance 2019.
89. Moseley et al. 2005; Nash 2019; Nash and deFrance 2019; Williams and Nash 2006; Williams et al. 2019.
90. Moseley et al. 2005.
91. Williams and Nash 2006.
92. Williams and Nash 2006:455.
93. Moseley et al. 2005:17,266, Fig. 4.
94. See Williams and Nash 2006 for further discussion regarding *Apu*.
95. Moseley et al. 2005.
96. Williams et al. 2019.
97. Moseley et al. 2005.
98. Moseley et al. 2005:17,267; Williams et al. 2019.
99. Williams et al. 2019.
100. Williams et al. 2019.
101. Moseley et al. 2005.
102. Moseley et al. 2005.
103. See D'Altroy 2015b.

104. Moore 1989:685.
105. Moore 1989.
106. Moseley 1992:256.
107. Moore 1989:685; Moseley 1992:256.
108. Moore and Mackey 2009:778–781.
109. Moore and Mackey 2009:781.
110. Moore 1989; Murra 1980:57.
111. 1989:685.
112. Netherly 1977:216–217; Moore 1989.
113. Moore 1989:685.
114. Swenson 2007.
115. Swenson 2007:65; Isbell 1997: Jackson 2004:314–316.
116. Jackson 2004; Swenson 2007:68.
117. Swenson 2007:69.
118. Swenson 2007:69–70.
119. Swenson 2007:65.
120. Dillehay and Kolata 2004; Cutright 2015:68.
121. Hayashida 2008; Moore 1989; Cutright 2015:80.
122. Cutright 2015:81.
123. Prieto 2011; Cutright 2015.
124. See Prieto 2011:120–121.
125. Prieto 2011:120.
126. Prieto 2011:114–115, Figs. 8 and 9.
127. Prieto 2011:118–119.
128. Prieto 2011:121.
129. Morris 1979:21–22.
130. Burger and Van der Merwe 1990; Gero 1990, 1992; Hastorf 1991; Hastorf and Johannessen 1993; Lau 2002; Moore 1989; Nash 2012; Shimada 1994.
131. See Stanish 2001.
132. Morris 1979:22; Valdez 2006.
133. Bray 2003a, 2003b; Duke 2010; Hastorf and Johannessen 1993; Jennings 2003; Jennings and Duke 2018; Moore 1989; Stanish 1997.
134. Hayashida 2019; D'Altroy 2015b:5; Jennings and Bowser 2009; Matos Mendieta and Barreiro 2015; Stanish 2001.
135. Espinoza 1987, as cited in Valdez 2006.
136. Ogburn 2004; see Dean 2010 for a full discussion on Inca stone.
137. Dean 2010:2.
138. Dean 2010:31–33, Fig. 9; Paternosto 1996:129.
139. Hastorf 2003:547.
140. Hastorf 2003:546–547.

141. Cummins 2002:110; Jennings and Bowser 2009:12.

142. D'Altroy and Hastorf 1992:265; Jennings 2005:252; Morris 1979:28, 32; Moseley et al. 2005; Shimada 1994:144, 169, 208, 222.

143. Bauer 1996:327; Coe 1994:206; Duke 2010:264; Hastorf and Johannessen 1993:120; Hastorf 2003:547; Jennings 2005:242; Jennings and Duke 2018:308.

144. Bray 2009; Jennings and Duke 2018; Nash 2010.

145. Murra 1960; Jennings 2005.

146. Bray 2003a:18–19; Hastorf and Johannessen 1993:118–119; Moore 1989:685; Morris 1979:32; Jennings 2005:243; Nash 2012.

147. D'Altroy and Earle 1992; Jennings and Duke 2018:315.

148. Murra 1980.

149. Rowe 1946:283.

150. Murra 1960.

151. Cobo 1990[1963]:172, as cited in Jennings and Duke 2018:315.

152. Jennings 2005:251–252.

153. Morris 1979:32; Hayashida 2019:51; Jennings 2005:244, 252.

154. Garcilaso de le Vega 1966:198; Weismantel 1991.

155. Morris 1979:32; Hayashida 2019:51.

156. Jennings 2005:27–249.

157. Allen 2009:37; Jennings and Duke 2018.

158. Bray 2003b.

159. Nash 2012:88.

160. Gagnon et al. 2015:175.

161. Hayashida 2019:51–52.

162. Hayashida 2019:51.

163. Moore 1989:685; Rostworowski 1989:279–280; Bernier 2010:35.

164. Matos Mendieta and Barreiro 2015:7; Hayashida 2019:51.

165. Hayashida 2019.

166. Arriaga 1968:41.

167. Kaulicke 2015:251.

168. Allen 2002:36; Bolin 1998:232; Jennings and Bowser 2009:10.

169. Bode 1990.

170. Jennings and Bowser 2009:14.

171. Reinhard 2007.

172. Zawaski and Malville 2007:22; Malville 2009.

173. Couture 2004; Salomon and Urioste 1991; Zawaski and Malville 2007:22–23.

174. Malville 2010.

175. Mackey 2010.

176. Kauliche 2015:252; Meddens et al. 2010; Nash 2012:88.
177. Meddens et al. 2010; Zuidema 2008:249, 261, 263–264; Pino Matos 2004; Zawaski and Malville 2007, 2010.
178. Zawaski and Malville 2007:22.
179. Malville 2010.
180. Malville 2010; Zuidema 2014.
181. Bauer and Stanish 2001; Dearborn et al. 1998.
182. Bauer and Stanish 2001: Malville 2010.
183. Malville 2010.
184. Salazar 2004; Malville 2010.
185. Malville 2010.
186. Wright and Valencia Zagarra 2004; Malville 2010.
187. Zawaski and Malville 2007:27.
188. Zawaski and Malville 2007:28.
189. Zuidema 1981.
190. Guaman Poma 1987 [1615]:86, as cited in Hastorf and Johannessen 1993.
191. Malville 2010.
192. Malville 2010; Zuidema 1981; Bauer 2004.
193. Bauer and Stanish 2001.
194. Zawaski and Malville 2007:29.
195. Zuidema 2014; Yaya 2015.
196. 1978 [1571].
197. Zuidema 2014.
198. Zuidema 2014; Dean 2010:138, 2015:219.
199. Zuidema 2014.
200. Dean 2007:506.
201. Orlove and Schmidt 1995; Perlov 2009.
202. Hayashida 2009; Moseley 1992; Murra 1973: Weismantel 1991.
203. Cutler and Cárdenas 1947; Jennings and Bowser 2009:13.
204. Bejarano 1950, as cited in Jennings and Bowser 2009:13.
205. Allen 2002; Hayashida 2008; Orlove and Schmidt 1995.
206. Allen 2009:41.
207. Allen 2009:34–38; Moseley 1992:49–65.
208. Allen 2009.
209. Allen 2009.
210. Allen 2009:34–38, 45.
211. Allen 2009: 39–41.
212. 1978.
213. Allen 2009:42–43.
214. Pevlov 2009:56.

215. 2009.
216. Pevlov 2009:58.
217. Pevlov 2009:56–62.
218. Pevlov 2009:64–69.
219. Bowser and Patton 2004; Uzendoski 2004.
220. Piperno 2011.
221. Bowser and Patton 2004; Dietler 2006:231–232; Haaland 2007.
222. Bowser and Patton 2004.
223. Bowser and Patton 2004:176.
224. Uzendoski 2004:883.
225. Uzendoski 2004:896.
226. 2004.
227. Uzendoski 2005:84.
228. Uzendoski 2004.
229. Uzendoski 2005:84.
230. Uzendoski 2004:896, 2005:18, 139.
231. Ericson and Baugh 1993; Hegmon et al. 2008; Washburn et al. 2011.
232. White 2017.
233. Smith 1989.
234. King et al. 2017.
235. See VanPool 2003; Dobkin de Rios 1976; Furst 1972; Harner 1973; Wilbert 1987; Winter 2000.
236. Myerhoff 1976; Schaefer and Furst 1996.
237. Moroukis 2010; Slotkin 1956; Schultes 1938.
238. Emerson 2003.
239. Robicsek 1978.
240. See VanPool 2003.
241. Hudson 1979.
242. Crown et al. 2012.
243. White 2017.
244. Hayashida 2008; Nicholson 1960.
245. Smalley and Blake 2003.
246. King et al. 2017; Whalen and Minnis 2009:171–176.
247. Bray 2009; Burger 2008; Murra 1980; Shimada 2001; Swenson 2006; Williams and Nash 2006.
248. Burger 2008, 2011.
249. Bray 2009; Shimada 1994, 2001; Swenson 2006.
250. Williams and Nash 2006; Moseley et al. 2005; Nash and deFrance 2019; Nash 2012.
251. Moore 1989.

252. Bray 2003a, 2003b; Hastorf and Johannessen 1993; Jennings and Duke 2018; Morris 1979; Valdez 2006.
253. Allen 2002; Hayashida 2008; Orlove and Schmidt 1995.

Chapter 7

1. Liu et al. 2018.
2. Liu et al. 2018.
3. Zhang et al. 2004.
4. Dietler and Herbich 2006; Homan 2004; McGovern 2009; Perruchini et al. 2018.
5. Homan 2004.
6. Munro 1996:69–71; Arthur 2014b.
7. Unger 2001:110.
8. Schaedel 1988, as described by Hayashida 2009:236; González Holguín 1989:18, 248, as described by Hayashida 2009:236.
9. Burger 2008.
10. Unger 2001:113.
11. Unger 2001:109.
12. Pearson 2017.
13. Griswold et al. 2018.
14. Bamforth 2002.
15. Bamforth 2002.
16. WHO 2017.
17. WHO 2020.
18. McGovern et al. 2005.
19. Smith and Lee 2018.
20. McHugh 2014; Sharma et al. 2010.
21. Armelagos 2000.
22. Cook et al. 1989.
23. Darby et al. 1977:548.
24. Edwardson 1952; Neve 1976.
25. Kennedy 1978.
26. See Liu et al. 2018.
27. Smith and Lee 2008; Joffe 1998; Pollack 1999; McGovern et al. 1999; Geller 1993; Pikirayi 2016: Brophy and Noble 2012; Rojo-Guerra et al. 2006; Dickson 1978; Dickson and Dickson 2002; Moseley et al. 2005; Dean 2010.

28. McGovern et al. 1999.
29. Munro 1996.
30. Arthur 2014b.
31. Pikirayi 2016.
32. Rojo-Guerra et al. 2006.
33. Dickson and Dickson 2002.
34. Dean 2010, 2015.
35. Samuel 1996b; Fontein 2006; McGovern et al. 1999; Dickson 1978; Hayashida 2019.
36. Gelb 1965; Michalowski 1994; Paulette 2020; Pollack 1999.
37. Wang et al. In press; Geller 1993.
38. Lerner 2018.
39. Burger 2008; Moore 1989; Morris 1979; Moseley 1992; Williams and Nash 2006.
40. Hoalst-Pullen et al. 2014.
41. Dineley 2006.
42. Cutler and Cardenas 1947; Nicholson 1960.
43. Geller 1993; McGovern et al. 1996, 1999, 2005; Wang et al. 2016, 2017.
44. Bouby et al. 2011; Rösch 2005; Wang et al. 2016.
45. Hines 2020.
46. Hoalst-Pullen et al. 2014.
47. Hoalst-Pullen et al. 2014.
48. Hoalst-Pullen et al. 2014:111.
49. Edmonds 2016; Hoalst-Pullen et al. 2014.
50. Masters Brewers Association of America 2020.
51. Hoalst-Pullen et al. 2014.
52. Manning 2018.
53. Manning 2018.
54. Olajire 2012; Dacek 2015; Mekonnen and Hoekstra 2010, 2011; Hoalst-Pullen et al. 2014.
55. Dacek 2015.
56. Glennon 2018.
57. Jones 2018; Sloan 2012.
58. Sloan 2012.
59. Yuengling Brewery 2015.
60. Yuengling Brewery 2018.
61. Yuengling Brewery 2015.
62. Jones 2018:15.
63. Hoalst-Pullen et al. 2014.
64. Hoalst-Pullen et al. 2014.

65. Sloan 2012.
66. Jennifer Sedillo 2020, personal communication.
67. Galanty 2016b; Hoalst-Pullen et al. 2014.
68. Brink 2019; Agehara 2020.
69. Brink 2019.
70. Agehara 2020.
71. Brink 2019; Agehara 2020.
72. Brink 2019.
73. Galanty 2016a.
74. Galanty 2016a.
75. Taylor et al. 2014.
76. Masters Brewers Association of America 2020.
77. See Harner 1973.
78. Wang 2021, personal communication.
79. Wang 2020, personal communication.
80. Wang 2020, personal communication.
81. Paulette 2020.
82. Ty Weaver, 2021, personal communication.
83. Lisa Kahn, 2021, personal communication.
84. Lisa Kahn, 2020, personal communication.
85. Lisa Kahn 2020, personal communication.
86. Kimberly G. Connor, 2020, personal communication.
87. Kimberly G. Connor, 2021, personal communication.
88. MacDonell 1910 [1669]:92.
89. Kimberly G. Connor, 2020, personal communication.
90. Jake Berman 2020, personal communication.
91. Jake Berman 2021, personal communication.
92. See also Hayashida 2008.
93. Arthur 2002, 2003, 2014a,2014b; Arthur and Weedman 2005.

Appendix: Ancient and Contemporary Beer Ingredient Tables

1. Liu et al. 2018.
2. McGovern et al. 2005.
3. Liu et al. 2019.
4. Wang et al. 2016; Liu et al. 2019.
5. Michel et al. 1993.

6. De Keersmaecker 1996.

7. Mohan and Sharma 1995:131.

8. McGovern et al. 1999.

9. Roy 1978.

10. Samuel 2000.

11. Attia et al. 2018; Friedman 2011:34.

12. Samuel 1996b.

13. Arthur 2014b.

14. Luning 2002.

15. Carlson 1990.

16. Dineley 2004:viii.

17. Dineley 2004:viii.

18. Noble and Brophy 2011.

19. Dineley and Dineley 2000:138; Wickham-Jones 1990.

20. McGovern 2013.

21. McGovern 2013.

22. See Behre 1999:37, Fig. 3 for a full list.

23. McGovern 2013.

24. De Keersmaecker 1996.

25. Nelson 2014.

26. See Behre 1999:38, 40–41, Fig. 4 and 6 for a full list.

27. Behre 1999:43.

28. Behre 1999:43.

29. Goldstein and Coleman 2004.

30. Goldstein et al. 2009:133.

31. Kennedy 1963:622; Irigoyen-Rascón 2015:232.

32. Irigoyen-Rascón 2015:186.

33. Irigoyen-Rascón 2015:204, 232.

34. Irigoyen-Rascón 2015:222.

35. Irigoyen-Rascón 2015:240–241.

36. Bye 1976:270, as cited in Irigoyen-Rascón 2015:333.

37. Goldstein and Coleman 2004:526.

38. Cutler and Cardenas 1947:40.

39. Cutler and Cardenas 1947:58.

Glossary

Bottom Fermentation uses yeast strains that work effectively between 41 and 50 degrees Fahrenheit, usually associated with lager beers.

Hops *Humulus lupulus* L. a plant from which brewers use the flowers or cones as a main ingredient of European-style beers, providing a bitter taste and giving beers their unique aroma and flavor. Hops have also helped to preserve beers for at least the past millennium.

International Bitterness Units (IBUs) an internationally agreed-on standard to measure bitterness in beer.

Kilning the final stage of the malting process, in which the brewer heats the germinated barley to dry it, resulting in a malty, biscuit-like flavor.

Lautering separating the wort from the spent grain.

Malt one of the four main ingredients of beer (i.e., water, hops, and yeast). The grain seeds are steeped/soaked in water to allow for germination, and then after germination are heated and dried in the kilning process.

Mash consists of ground cereal grains and water, resulting in a porridge-like mixture.

Milling the physical crushing of malt kernels, usually with a ground stone by Indigenous brewers to prepare the grain for mashing.

Phytolith a microscopic silica found in some plant tissues that survive after the plant has decayed, allowing archaeobotanists to determine the type of plants people were processing and potentially eating in the past.

Sparge the spraying of hot liquor (brewing water) onto the mash to rinse out the residual sugars.

Straining separates the solids in the mash (the spent grain) from the liquids (the wort that becomes beer).

Top Fermentation uses yeast strains that work effectively above 60.8 degrees Fahrenheit, usually associated with ale beers.

Tun a vessel in which the brewers mix the ground mash (grist) with the temperature-controlled water.

Wort the watery mixture of ground malt before fermentation.

Wort Boiling The wort is boiled with hops, giving the wort its rich, dark color and reducing the pH of the wort, allowing fermentation to begin.

Wort Separation The grain husks are separated from the mash during the lautering process.

Yeast a variety of unicellular fungi that allows the wort to turn into beer by consuming wort sugars and in return giving off alcohol, carbon dioxide, and a range of flavors.

Yeast Pitching the process of adding yeast to the wort to begin the fermentation process.

References

Abbink, Jon. 2002. Drinking, Prestige, and Power: Alcohol and Cultural Hegemony in Maji, Southern Ethiopia. In *Alcohol in Africa: Mixing Business, Pleasure and Politics*, edited by Deborah Fahy Bryceson, pp. 161–178. Portsmouth, NH: Heinemann.

Agehara, Shinsuke. 2020. Florida Hops Show Potential. *Vegetable and Specialty Crop News Magazine*. https://vscnews.com/florida-hops-show-potential/?fbclid=IwAR039FBrt5kxn3y4HzLuS88b4xxoPUOuPiDTqnXFibCuTyLk3cLQ_DZQub0.

Aka, Solange, Georgette Konan, Gilbert Fokou, Koffi Marcellin Dje, and Bassirou Bonfoh. 2014. Review of African Traditional Cereal Beverages. *American Journal of Research Communication* 2(5):103–153.

Albarella U., and D. Serjeantson. 2002. A Passion for Pork: Butchery and Cooking at the British Neolithic site of Durrington Wall. In *Consuming Passions and Patterns of Consumption*, edited by P.T. Miracle and N. Milner, pp. 33–49. Cambridge: MacDonald Institute.

Alexandersen, V., P. Bennike, L. Hvass, and Nielsen K.H. Stærmose. 1981. Egtvedpigen—nye undersøgelser. *Aarbøger for Nordisk Oldkyndighed og Historie*, pp.17–46.

Ali, S.S. El-Ahmadhy, N. Ayuob, A.N. Singab. 2015. Phytochemicals of Markhamia Species (Bignoniaceae) and Their Therapeutic Value: A Review. *European Journal of Medicinal Plants* 6:124–142.

Allchin, B., and R. Allchin. 1982. *The Rise of Civilization in India and Pakistan*. Cambridge: Cambridge University Press.

Allchin, F.R. 1979. The Ancient Home of Distillation? *Man* 14(1):55–63.

Allen, Catherine J. 2002. *The Hold Life Has: Coca and Cultural Identity in an Andean Community*. 2nd ed. Washington, DC: Smithsonian Institution Press.

Allen, Catherine J. 2009. Let's Drink Together, My Dear! Persistent Ceremonies in a Changing Community. In *Drink, Power and Society in the Andes*, edited by J. Jennings and B. Bowser, pp. 28–48. Gainesville: University Press of Florida.

Alm, Torbjorn. 2003. Ales, Beer and Other Viking Beverages—Some Notes Based on Norwegian Ethnobotany. *Yearbook of the Heather Society* 2003:37–44.

Alm, Torbjorn, and Brita Elvevåg. 2013. Ergotism in Norway. Part 1: The Symptoms and Their Interpretation from the Late Iron Age to the Seventeenth Century. *History of Psychiatry* 24(1):15–33.

Anderson, Karen. 2009. Tiwanau Influence on Local Drinking Patterns in Cochabamba, Bolivia. In *Drink, Power, and Society in the Andes*, edited by Justin Jennings and Brenda J. Bowser, pp. 167–199. Gainesville: University Press of Florida.

Applequist, Wendy L., and Daniel E. Moerman. 2011. Yarrow (Achillea millefolium L.): A Neglected Panacea? A Review of Ethnobotany, Bioactivity, and Biomedical Research. *Economic Botany* 65(2):209–225.

Armelagos, George J. 2000. Take Two Beers and Call Me in 1,600 Years: Ancient Nubians and Egyptians Had a Way with Antibiotics. *Natural History* 5:50–53.

Armstrong, David E. 1993. *Drinking with the Dead: Alcohol and Altered States in Ancestor Veneration Rituals of Zhou Dynasty China and Iron Age Palestine.* MA thesis, York University, Toronto, Ontario, Canada.

Arnold, Bettina. 1999. "Drinking the Feast": Alcohol and the Legitimation of Power in Celtic Europe. *Cambridge Archaeological Journal* 9(1):71–93.

Arnold, J.P. 1911. *Origin and History of Beer and Brewing from Prehistoric Times to the Beginning of Brewing Science and Technology: A Critical Essay.* Chicago: Alumni Association of the Wall-Henius Institute of Fermentology.

Arranz-Otaegui, Amia, Lara G. Carretero, Monica N. Ramsey, Dorian Q. Fuller, and Tobias Richter. 2018. Archaeobotanical Evidence Reveals the Origins of Bread 14,400 Years Ago in Northeastern Jordan. *Proceedings of the National Academy of Sciences of the United States of America* 115(31): 7925–7930.

Arriaga, Pedro J. de. 1968 [1621]. *The Extirpation of Idolatry in Peru.* Lexington: University of Kentucky Press.

Arthur, John W. 2002. Pottery Use-Alteration as an Indicator of Socioeconomic Status: An Ethnoarchaeological Study of the Gamo of Ethiopia. *Journal of Archaeological Method and Theory* 9:331–355.

Arthur, John W. 2003. Brewing Beer: Status, Wealth, and Ceramic Use-Alteration among the Gamo of Southwestern Ethiopia. *World Archaeology* 34:516–528.

Arthur, John W. 2006. *Living with Pottery: Ethnoarchaeology among the Gamo of Southwest Ethiopia*, Foundations of Archaeological Inquiry. Salt Lake City: The University of Utah Press.

Arthur, John W. 2009. Understanding Household Population through Ceramic Assemblage Formation. Ceramic Ethnoarchaeology among the Gamo of Southwestern Ethiopia. *American Antiquity* 74(1):31–48.

Arthur, John W. 2013. Transforming Clay: Gamo Caste, Gender, and Pottery of Southwestern Ethiopia. *African Study Monographs, Supplementary Issue* 46:5–25.

Arthur, John W. 2014a. Culinary Crafts and Foods in Southwestern Ethiopia: An Ethnoarchaeological Study of Gamo Groundstones and Pottery. *African Archaeological Review 31*(2):131–168.

Arthur, John W. 2014b. Beer through the Ages: The Role of Beer in Shaping Our Past and Current Worlds. *Anthropology Now 6*(2):1–11.

Arthur, John W. 2014c. Pottery Uniformity in a Stratified Society: An Ethnoarchaeological Perspective from the Gamo of Southwestern Ethiopia. *Journal of Anthropological Archaeology 35*:106–116.

Arthur, John W. 2019. Built to Last: James Skibo's Contribution to Archaeological Theory and Pottery Use-Alteration in Relation to the Gamo of Southwestern Ethiopia. *The Wisconsin Archaeologist* in a special issue on the contributions of James Skibo. Editors Fernanda Neubauer and Michael J. Schaefer. *100*(1):91–104.

Arthur, John W., and Kathryn Weedman. 2005. *Ethnoarchaeology, Handbook of Archaeological Methods*, edited by Herbert Machner, pp. 216–269. Lanham, MD: AltaMira Press.

Arthur, John W., et al. 2019. The Transition from Hunting and Gathering to Food Production in the Gamo Highlands of Southern Ethiopia. *African Archaeological Review 36*(1):5–65.

Atalay, Sonya, and Christine A. Hastorf. 2006. Food, Meals, and Daily Activities: Food Habitus at Neolithic Catalhoyuk. *American Antiquity 71*(2):283–319.

Attia, Elshafaey A.E., Elena Marinova, Ahmed G. Fahmy, and Masahiro Baba. 2018. Archaeobotanical Studies from Hierakonpolis: Evidence for Food Processing during the Predynastic Period in Egypt. In *Plants and People in the African Past: Progress in African Archaeobotany*, edited by A.M. Mercuri, Catherine D'Andrea, Rita Fornaciari, and Alexa Höhn, pp. 76–89. Cham, Switzerland: Springer.

Aurenche, O., and S.K. Koslowski. 2001. Le Croissant fertile et le "Triangle d'or." In *Études Mésopotamiennes: Receuil de textes offert à Jean-Louis Huot*, edited by C. Breniquet and C. Kepinski, pp. 33–43. Éditions Recherche sur les Civilisations, Paris.

Baba, Masahiro, and Renee. F. Friedman. 2016. Recent excavations at HK11C, Hierakonpolis. In *Egypt at Its Origins 4: Proceedings of the Fourth International Conference "Origin of the State. Predynastic and Early Dynastic Egypt," New York, July 26–30, 2011*, edited by M.D. Adams and collaborators B. Midant-Reynes, E.M. Ryan, and Y. Tristant, pp. 179–205. Bristol, CT: Orientalia Lovaniensia Analecta 252.

Baba, Masahiro. In press. *Ceramic Assemblages from HK11C at Hierakonpolis: Specialization Examined*. In *Egypt at its Origins 6: Proceedings of the Sixth International Conference on Origin of the State, Predynastic and Early Dynastic Egypt, Vienna, 10th–15th September 2017*. Leuven, Belgium: Peeters.

Ballard, Mary W. 2012. King Midas' Textiles and His Golden Touch. In *The Archaeology of Phrygian Gordion, Royal City of Midas*, edited by C.B. Rose, 165–170. Philadelphia: University of Pennsylvania Museum.

Bamforth, Charles W. 2002. Nutritional Aspects of Beer—A Review. *Nutrition Research 22*:227–237.

Bamforth, Charles W. 2004. *Beer: Health and Nutrition*. Oxford: Blackwell.

Banning, E.B. 1998. The Neolithic Period: Triumphs of Architecture, Agriculture, and Art. *Near Eastern Archaeology 61*:188–237.

Banning, E.B. 2011. So Fair a House: Gobekli Tepe and the Identification of Temples in the Pre-pottery Neolithic of the Near East. *Current Anthropology 52*(5):619–660.

Barclay, G.J. 1983. Sites of the Third Millennium BC to the First Millennium AD at North Mains, Strathallan, Perthshire. *Proceedings of the Society of Antiquaries of Scotland 113*: 122–281.

Barclay, G.J., and C.J. Russell-White. 1993. Excavations in the Ceremonial Complex of the Fourth to the Second Millennium BC at Balfarg/Balbirnie, Glenrothes, Fife. *Proceedings of the Society of Antiquaries of Scotland 123*:43–210.

Baron, Stanley (Wade). 1972. *Brewed in America: A History of Beer and Ale in the United States*. New York: Arno Press.

Barth, F. 1967. Economic Spheres in Darfur. In *Themes in Economic Anthropology*, edited by R. Firth, pp. 149–174. London: Travistock.

Bar-Yosef, Ofer, and David Alon. 1988. Nahal Hemar Cave. Jerusalem: Atiqot, English Series, volume 18.

Bauer, Brian S. 1996. Legitimization of the State in Inca Myth and Ritual. *American Anthropologist 98*(2):327–337.

Bauer, Brian S. 2004. *Ancient Cuzco: Heartland of the Inca*. Austin: University of Texas Press.

Bauer, Brian S., and Charles Stanish. 2001. *Ritual and Pilgrimage in the Ancient Andes: The Islands of the Sun and the Moon*. Austin: University of Texas Press.

Behre, Karl-Ernst. 1999. The History of Beer Additives in Europe—A Review. *Vegetation History and Archaeology 8*(1–2):35–48.

Bejarano, Jorge. 1950. *La Derrota de un Vicio: Origen e Historia de la Chicha*. Bogotá: Editorial Iqueima.

Bender, Barbara. 1978. Gatherer-Hunter to Farmer: A Social Perspective. *World Archaeology 10*(2): 204–222.

Bennett, Judith M. 1991. Misogyny, Popular Culture, and Women's Work. *History Workshop 31*:166–188.

Benz, Marion. 2010. The Principle of Sharing—An Introduction. In *Principle of Sharing, Segregation and Construction of Social Identities at the Transition from Foraging to Farming. Studies in Early Near Eastern Production,*

Subsistence, and Environment, edited by Marion Benz, pp.1–18. Berlin: Ex Oriente.

Bernier, Hélène. 2010. Craft Specialists at Moche: Organization, Affiliations, and Identities. *Latin American Antiquity 21*(1):22–43.

Berns, M.C. 1988. Ga'anda Scarification. In *Marks of Civilization,* edited by A. Rubin, pp. 51–76. Los Angeles: Museum of Cultural History, University of California.

Bertman, S. 2003. *Handbook to Life in Ancient Mesopotamia.* Oxford: Oxford University Press.

Blasco, A., M. Edo, M.J. Villalba, R. Buxó, J. Juan-Tresserras, and M. Saña Seguí 1999. Del Cardial al Postcardial en la Cueva de Can Sadurní (Begues, Barcelona). Primeros Datos sobre su Secuencia Estratigráfica, Paleoeconómica y Ambiental. In *Actes del II Congrés del Neolític a la Península Ibèrica,* edited by J. Bernabeu and T. Orozco, Saguntum Supplement 2, pp. 59–68. València: Universitat de València.

Bode, Barbara. 1990. *No Bells to Toll: Destruction and Creation in the Andes.* New York: Scribners.

Boer, Roland. 2018. From Horse Kissing to Beastly Emissions: Paraphilias in the Ancient Near East. In *Sex in Antiquity: Exploring Gender and Sexuality in the Ancient World,* edited by Mark Masterson, Nancy Sorkin Rabinowitz, and James Robson, pp. 67–79. New York: Routledge.

Boesche, R. 2003. *The First Great Political Realist: Kautilya and His Arthashastra.* Lanham, MD: Lexington Books.

Bolin, I. 1998. *Rituals of Respect.* Austin: University of Texas Press.

Bouby, Laurent, Philippe Boissinot, and Philippe Marinval. 2011. Never Mind the Bottle. Archaeobotanical Evidence of Beer-Brewing in Mediterranean France and the Consumption of Alcoholic Beverages during the 5th Century BC. *Human Ecology 29*:351–360.

Bowser, Brenda J., and John Q. Patton. 2004. Domestic Spaces as Public Places: An Ethnoarchaeological Case Study of Houses, Gender, and Politics in the Ecuadorian Amazon. *Journal of Archaeological Method and Theory 11*(2):157–181.

Boyadjian, Célia Helena C., Sabine Eggers, and Karl Reinhard. 2007. Dental Wash: A Problematic Method for Extracting Microfossils from Teeth. *Journal of Archaeological Science 34*(10):1622–1628.

Braidwood, R.J., J.D. Sauer, J. Helbaek, P.C. Magledorf, H.C. Cutler, C.S. Coon, R. Linton, J. Steward, and A.L. Oprenheim. 1953. Symposium: Did Man Once Live on Beer Alone? *American Anthropologist 53*(4):515–526.

Bray, Tamara L. 2003a. Inca Pottery as Culinary Equipment: Food, Feasting, and Gender in Imperial Design. *Latin American Antiquity 14*(1):3–28.

Bray, Tamara L. 2003b. To Dine Splendidly: Imperial Pottery, Commensal Politics, and the Inca State. In *The Archaeology and Politics of Food and*

Feasting in Early States and Empires, edited by Tamara L. Bray, pp. 93–142. New York: Kluwer Academic/Plenum.

Bray, Tamara L. 2009. The Role of Chicha in Inca State Expansion. In *Drink, Power, and Society in the Andes*, edited by Justin Jennings and Brenda J. Bowser, pp. 108–132. Gainesville: University Press of Florida.

Brettell, Rhea, Janet Montgomery, and Jane Evans. 2012. Brewing and Stewing: The Effect of Culturally Mediated Behaviour on the Oxygen Isotope Composition of Ingested Fluids and the Implications for Human Provenience Studies. *Journal of Analytical Atomic Spectrometry* 27(5):778–785.

Briggs, D.E., C. Boulton, P. Brooks, and R. Stevens. 2004. *Brewing: Science and Practice*. Cambridge: Woodhead.

Brink, Graham. 2019. A Hop and a Prayer—Scientists Cultivate Key Beer Ingredient in Hillsborough County. *Tampa Bay Times*, December 14. https://www.tampabay.com/news/business/2019/12/14/a-hop-and-a-pra yer-scientists-cultivate-key-beer-ingredient-in-hillsborough-county/.

Brombacher, C., S. Jacomet, and M. Kühn. 1997. Mittelalterliche Kulturpflanzen aus der Schweiz und Liechtenstein: eine Über sicht der archäobotanischen Nachweise. In *Papers of the Medieval Europe Brugge 1997. Conference volume 90*, edited by G. de Boe and F. Verhaeghe, pp. 95–111. I.A.P, Rapporten 9.

Brophy, Kenneth, and Gordon Noble. 2009. *Forteviot, Perthshire, 2009: Excavations of a Henge and Cist Burial. Data tructure and Interim Report Prepared for the Strathearn Environs and Royal Forteviot (SERF) project*, University of Glasgow.

Brophy, K., and G. Noble. 2012. Henging, Mounding and Blocking: The Forteviot Henge Group. In *Enclosing the Neolithic: Recent Studies in Britain and Europe*, edited by A. Gibson, pp. 21–35. Oxford: British Archaeological Reports S2440.

Bryant, A.T. 1970. *The Zulu People as They Were before the White Man Came.* Shuter and Shooter, 1948; reprinted, Negro Universities Press, New York.

Bryceson, D.F. 2002. *Alcohol in Africa: Mixing Business, Pleasure, and Politics.* Portsmouth, NH: Heinemann.

Budge, E.A. Wallis. 2012. *Osiris and the Egyptian Resurrection.* New York: Dover Publications.

Burger, Richard L. 1984. *The Prehistoric Occupation of Chavín de Huántar, Peru.* University of California Publications in Anthropology 14. Berkeley: University of California Press.

Burger, Richard L. 1992. *Chavin and the Origins of Andean Civilization.* London: Thames and Hudson.

Burger, Richard L. 2008. Chavín de Huántar and Its Sphere of Influence. In *Handbook of South American Archaeology*, edited by Helaine Silverman and William H. Isbell, pp. 681–702. New York: Springer.

Burger, Richard L. 2011. What Kind of Hallucinogenic Snuff Was Used at Chavín de Huántar? An Iconographic Identification. *Ñawpa Pacha Journal of Andean Archaeology* 31(2): 123–140.

Burger, Richard L., and Nikolaas J. Van der Merwe. 1990. Maize and the Origin of Highland Chavín Civilization: An Isotopic Perspective. *American Anthropologist* 92(1):86–96.

Butters, Luis Jaime Castillo, and Santiago Uceda Castillo. 2008. The Mochicas. In *Handbook of South American Archaeology*, edited by Helaine Silverman and William H. Isbell, pp. 707–730. New York: Springer.

Bye, Robert A. 1976. *Ethnoecology of the Tarahumara of Chihuahua, Mexico*. PhD dissertation, Harvard University.

Byock, J. 2001. *Viking Age Iceland*. London: Penguin.

Camps, G. 1974. *Les Civilisations préhistoriques de l'Afrique du Nord et du Sahara*. Paris: Doin.

Carlson, Robert G. 1990. Banana Beer, Reciprocity, and Ancestor Propitiation among the Haya of Bukoba, Tanzania. *Ethnology* 29(4):297–311.

Carod-Artal, F.J. 2015. Hallucinogenic Drugs in Pre-Columbian Mesoamerican Cultures. *Neurología*, 30(1):42–49.

Centeno Martín, Marcos P. 2018. Contextualising N.G. Munro's Filming of the Ainu Bear Ceremony. *Japan Society Proceedings*:90–106.

Cernea, S. 2009. Preface: An Original Contribution to Country-Wide Displacement Analysis. In *Moving People in Ethiopia: Development, Displacement and the State*, edited by Alula Pankhurst and Francois Piguet, pp. xxv–xxx. Oxford: James Currey.

Chang, K.C. 1980. *Shang Civilization*. New Haven, CT: Yale University Press.

Chang, Kwang-Chih. 1999. China on the Eve of the Historical Period. In *The Cambridge History of Ancient China: From the Origins of Civilization to 221 BC*, edited by Michael Loewe and Edward L. Shaughnessy, pp. 37–73. New York: Cambridge University Press.

Chi, Z., and H. Hung. 2013. Jiahu I: Earliest Farmers beyond the Yangtze River. *Antiquity* 87:46–63.

Chung, Saehyang P. 2008. Flying Cranes and Drifting Cloud Motifs on Koryo Celadons: Their Origin and Dissemination. *Acta Koreana* 11(2):115–139.

Cilurzo, Vincent. 2012. Sour Beer. In *Oxford Companion to Beer*, edited by Garrett Oliver, pp. 743–745. New York: Oxford University Press.

Civil, Miguel. 1964. A Hymn to the Beer Goddess and a Drinking Song. In *From the Workshop of the Chicago Assyrian Dictionary: Studies Presented to A. Leo Oppenheim, AS 27*. XXX Edition, pp. 67–89. Chicago: University of Chicago Press.

Cobo, Fray B. 1990 [1653]. *Inca Religion and Customs*. Austin: University of Texas Press.

Coe, Sophie D. 1994. *America's First Cuisines*. Austin: University of Texas Press.

Coleman, John E., et al. 2017. The Environment and Interactions of Neolithic Halai. In *Communities, Landscapes, and Interaction in Neolithic Greece: Proceedings of the International Conference, Rethymno 29–30 May, 2015*, edited by A. Sarris, E. Kalogiropoulou, T. Kalayci, and L. Karimali, pp. 97–125. Ann Arbor, MI: International Monographs in Prehistory, Archaeological Series 20.

Colson, E., and T. Scudder. 1988. *For Prayer and Profit: The Ritual, Economic, and Social Importance of Beer in Gwembe District, Zambia, 1950–1982*. Stanford, CA: Stanford University Press.

Cook, M., E.L. Molto, and C. Anderson. 1989. Flourochrome Labelling in Roman Period Skeletons from Dakhleh Oaisis, Egypt. *American Journal of Physical Anthropology* 80:137–143.

Cool, H.E.M. 2006. *Eating and Drinking in Roman Britain*. Cambridge: Cambridge University Press.

Cooper, J.S. 2016. The Job of Sex: The Social and Economic Role of Prostitutes in Ancient Mesopotamia. In *The Role of Women in Work and Society in the Ancient Near East*, edited by B. Lion and C. Michel, pp. 209–227. Boston: De Gruyter.

Cornelius, Izak. 1989. The Garden in the Iconography of the Ancient Near East: A Study of Selected Material from Egypt. *Journal of Semitics* 1(2):204–228.

Costin, Cathy. 2001. *Production and Exchange of Ceramics*. In *Empire and Domestic Economy*, edited by Terrance D'Altroy and Christine Hastorf, pp. 203–242. New York: Kluwer Academic/Plenum.

Couture, Nicole C. 2004. Monumental Space, Courtly Style, and Elite Life at Tiwanaku. In *Tiwanakii. Ancestors of the Inca*, edited by Margaret Young-Sanchez, pp. 126–149. Albuquerque: University of New Mexico Press.

Craig, Oliver E., et al. 2015. Feeding Stonehenge: Cuisine and Consumption at the Late Neolithic Site of Durrington Walls. *Antiquity* 89:1096–1109.

Crewe, Lindy, and Ian Hill. 2012. Finding Beer in the Archaeological Record: A Case Study of Kissonerga-Skalia on Bronze Age Cyprus. *Levant* 44(2):205–237.

Crosswhite, F.C. 1982. Corn (Zea mays) in Relation to Wild Relatives. *Desert Plants* 3:193–202.

Croucher, Karina. 2012. *Death and Dying in the Neolithic Near East*. Oxford: Oxford University Press.

Crown, Patricia L., Thomas E. Emerson, Jiyan Gu, W. Jeffrey Hurst, Timothy R. Pauketat, and Timothy Ward. 2012. Ritual Black Drink Consumption at Cahokia. *Proceedings of the National Academy of Sciences* 109(35):13944–13,949.

Crush, J.S. 1992. The Construction of Compound Authority: Drinking at Havelock, 1938–1944. In *Liquor and Labor in Southern Africa*, edited by Jonathan Crush and Charles Ambler, pp. 367–394. Athens: Ohio University Press.

Cummins, Thomas B.F. 2002. *Toasts with the Inca: Andean Abstraction and Colonial Images on Quero Vessels.* Ann Arbor: University of Michigan Press.

Cunningham, J.J. 2013. *The Independence of Ethnoarchaeology.* In *Human Expeditions: Inspired by Bruce G. Trigger,* edited by S. Chrisomalis and A. Costopoulos, pp. 51–72. Toronto: University of Toronto Press.

Curry, Andrew. 2008. *Gobekli Tepe: The World's First Temple? Smithsonian* November:1–6.

Curtis, John. 1838. *Shipwreck of the Stirling Castle, George Virtue.* London: Ivy Lane.

Cutler, Hugh C., and Martin Cardenas. 1947. Chicha, a Native South American Beer. *Botanical Museum Leaflets, Harvard University* 13(3):33–60.

Cutright, Robyn E. 2015. Eating Empire in the Jequetepeque: A Local View of Chimú Expansion on the North Coast of Peru. *Latin American Antiquity* 26(1):64–86.

Dacek, Tera. 2015. The Craft Beer Industry & Sustainability: More Than a Pint-Size Impact. *University of Vermont Out Reach,* The University of Vermont Continuing and Distance Education. https://learn.uvm.edu/blog/blog-business/craft-beer-industry-sustainability.

Daines, Alison. 2008. Egyptian Gardens. *Studia Antiqua* 6(1):15–25.

D'Altroy, Terence N. 2015a. Funding the Inka Empire. In *The Inka Empire: A Multidisciplinary Approach,* edited by I. Shimada, pp. 67–82. Austin: University of Texas Press.

D'Altroy, Terence N. 2015b. *The Incas.* 2nd ed. Oxford: Blackwell.

D'Altroy, Terence N., and Timothy K. Earle. 1992. Inka Storage Facilities in the Upper Mantaro Valley, Peru. In *Inka Storage Systems,* edited by Terry Y. LeVine, pp. 176–205. Norman: University of Oklahoma Press.

D'Altroy, Terence N., and Christine A. Hastorf. 1992. The Architecture and the Contents of Inka Storehouses in the Xauta Region. In *Inka Storage Systems,* edited by Terry LeVine, pp. 259–286. Norman: University of Oklahoma Press.

D'Altroy, Terence N., and Christine A. Hastorf. 2001. *Empire and Domestic Economy.* New York: Kluwer Academic/Plenum Publishers.

Damerow, Peter. 2012. Sumerian Beer: The Origins of Brewing Technology in Ancient Mesopotamia. *Cuneiform Digital Library Journal* 2:1–20.

Daniels, Ray. 2012. Bamberg, Germany. In *The Oxford Companion to Beer,* edited by Garrett Oliver, p. 83. Oxford: Oxford University Press.

Darby, W.J., P. Ghalioungui, and L. Grivetti. 1977. *Food: The Gift of Osiris, Volume 2.* London: Academic Press.

Darbyshire, Gareth, and Gabriel H. Pizzorno. 2009. Gordion in History. *Expedition: The Magazine of the University of Pennsylvania* 51(2):11–22.

Dean, Carolyn. 2007. The Inka Married the Earth: Integrated Outcrops and the Making of Place. *The Art Bulletin* 89(3):502–518.

Dean, Carolyn J. 2010. *A Culture of Stone: Inka Perspectives on Rock*. Durham, NC: Duke University Press.

Dean, Carolyn. 2015. *Men Who Would Be Rocks: The Inka Wank'a*. In *The Archaeology of Wak'as: Exploration of the Sacred in the Pre-Columbian Andes*. edited by Tamara L. Bray, pp. 213–238. Boulder: University Press of Colorado.

Dearborn, David S.P., Matthew T. Seddon, and Brian S. Bauer. 1998. The Sanctuary of Titicaca: Where the Sun Returns to Earth. *Latin American Antiquity* 9(3):240–258.

de Garine, I. 1996. Food and the Status Quest in Five African Cultures. In *Food and the Status Quest: An Interdisciplinary Perspective*, edited by P. Wiessner and W. Schiefenhövel, pp. 193–217. Providence, RI: Berghahn Books.

De Graef, K. 2018. In Taberna Quando Sumus. On Taverns, Nadītum Women, and the Gagûm in Old Babylonian Sippar. In *Gender and Methodology in the Ancient Near East: Approaches from Assyriology and Beyond*, edited by S.L. Bundin, M. Cifarelli, A. Garcia-Ventura, and A.M. Alba, pp. 77–115. Barcelona: Edicions de la Universitat de Barcelona.

De Keersmaecker, Jacques. 1996. The Mystery of Lambic Beer. *Scientific American*, August, 74–80.

Dell, Melissa. 2010. The Persistent Effects of Peru's Mining Mita. *Econometrica* 78(6):1863–1903.

DeLyser, D.Y. , and W.J. Kasper. 1994. Hopped Beer: The Case for Cultivation. *Economic Botany* 48(2):166–170.

Deuraseh, Nurdeen. 2003. Is Imbibing Al-Khamr (intoxicating drink) for Medical Purposes Permissible by Islamic law? *Arab Law Quarterly* 18:256–360.

Dickson, C., and J.H. Dickson. 2002. *Plants and People in Ancient Scotland*. Charleston, SC: Arcadia.

Dickson, J.H. 1978. Bronze Age Mead. *Antiquity* 52:108–113.

Dietler, Michael. 1990. Driven by Drink: The Role of Drinking in the Political Economy and the Case of Early Iron Age France. *Journal of Anthropological Archaeology* 9:352–406.

Dietler, Michael. 2001. Theorizing the Feast: Rituals of Consumption, Commensal Politics, and Power in African Contexts. In *Feasts: Archaeological and Ethnographic Perspectives on Food, Politics, and Power*, edited by Michael Dietler and Brian Hayden, pp. 65–114. Washington, DC: Smithsonian Institution Press.

Dietler, Michael. 2006. Alcohol: Anthropological/Archaeological Perspectives. *Annual Review of Anthropology* 35:229–249.

Dietler, M., and Hayden, B. (Eds). 2001. *Feasts: Archaeological and Ethnographic Perspectives on Food, Politics, and Power.* Washington, DC: Smithsonian Institution Press.

Dietler, Michael, and Ingrid Herbich. 2001. Feasts and Labor Mobilization: Dissecting a Fundamental Economic Practice. In *Feasts: Archaeological and Ethnographic Perspectives on Food, Politics, and Power,* edited by Michael Dietler and Brian Hayden, pp. 240–64. Washington, DC: Smithsonian Press.

Dietler, Michael, and Ingrid Herbich. 2006. Liquid Material Culture: Following the Flow of Beer among the Luo of Kenya. In *Grundlegungen. Beiträge zur europäischen und afrikanischen Archäologie für Manfred K.H. Eggert,* edited by Hans-Peter Wotzka, pp. 395–407. Tübingen, Germany: Francke.

Dietrich, L., et al. 2019. Cereal Processing at Early Neolithic Göbekli Tepe, Southeastern Turkey. *PLOS ONE 14*(5), e0215214. DOI: 10.1371/journal.pone.0215214.

Dietrich, Laura, et al. 2020. Investigating the Function of Pre-pottery Neolithic Stone Troughs from Göbekli Tepe—An Integrated Approach. *Journal of Archaeological Science: Reports 34,* Part A:1–20.

Dietrich, Oliver, Manfred Heun, Jens Notroff, Klaus Schmidt, and Martin Zarnkow. 2012. The Role of Cult and Feasting in the Emergence of Neolithic Communities. New Evidence from Gobekli Tepe, South-Eastern Turkey. *Antiquity 86*:674–95.

Dillehay, Tom D., and Alan L. Kolata. 2004. Long-Term Human Response to Uncertain Environmental Conditions in the Andes. *Proceedings of the National Academy of Sciences 101*(12):4325–4330.

Dineley, Merryn. 2004. *Barley, Malt, and Ale in the Neolithic.* BAR International Series. Oxford: Archaeopress.

Dineley, Merryn. 2006. The Use of Spent Grain as Cattle Feed in the Neolithic. In *Animals in the Neolithic of Britain and Europe,* edited by D. Serjeantson and D. Field, pp. 56–62. Oxford: Oxbow Press.

Dineley, Merryn. 2015. The Craft of the Maltster. In *Food and Drink in Archaeology 4,* edited by Wendy Howard, Kirsten Bedigan, and Ben Jervis, pp. 63–71. London: Prospect Books.

Dineley, M., and G. Dineley. 2000. Neolithic Ale: Barley as a Source of Malt Sugars for Fermentation. In *Plants in Neolithic Britain and Beyond,* edited by A.S. Fairbairn, pp. 137–154. Oxford: Oxford University Press.

Dirar, H. A. 1993. *The Indigenous Fermented Foods of the Sudan: A Study of African Fermented Foods and Nutrition.* Wallingford, UK: CAB International.

Djien, K.S. 1982. Indigenous Fermented Foods. In *Economic Microbiology, Vol. 7, Fermented Foods,* edited by A.H. Rose, pp. 15–38. London: Academic Press.

Dobkin de Rios, M. 1976. *The Wilderness of Mind: Cross-Cultural Perspective.* Beverly Hills: Sage.

Donham, D.L. 1999. *History, Power, and Ideology: Central Issues in Marxism and Anthropology*. Berkeley: University of California Press.

Dornbusch, Horst, and Garrett Oliver. 2012. Porter. In *The Oxford Companion to Beer*, edited by G. Oliver, pp. 660–664. New York: Oxford University Press.

Douglas, John E., and A.C. MacWilliams. 2015. Society and Polity in the Wider Casas Grandes Region. In *Ancient Paquimé and the Casas Grandes World*, edited by Paul E. Minnis and Michael E. Whalen, pp. 126–148. Amerind Studies in Anthropology. Tucson: University of Arizona Press.

Duke, Guy S. 2010. Continuity, Cultural Dynamics, and Alcohol: The Reinterpretation of Identity through Chicha in the Andes. In *Identity Crisis: Archaeological Perspectives on Social Identity*, edited by Lindsey Amundsen-Meyer, Nicole Engel, and Sean Pickering, pp. 263–272. Calgary: Chacmool Archaeology Association.

Dunne, Julie, et al. 2012. First Dairying in Green Saharan Africa in the Fifth Millennium BC. *Nature* 486(7403):390–394.

Dunne, Julie, Anna Maria Mercuri, Richard P. Evershed, Silvia Bruni, and Savino di Lernia. 2016. Earliest Direct Evidence of Plant Processing in Prehistoric Saharan Pottery. *Nature Plants* 3(16,194):1–6.

Eaton, Brian. 2018. An Overview of Brewing. In *Handbook of Brewing*, 3rd ed., edited by Graham G. Stewart, Inge Russell, and Annae Anstruther, pp. 53–66. Boca Raton, FL: CRC Press, Taylor & Frances Group.

Edgerton, William F. 1951. The Strikes in Ramses III's Twenty-Ninth Year. *Journal of Near Eastern Studies* 10(3):137–45.

Edmonds, David. 2016. Before the Taps Run Dry: Incentivizing Water Sustainability in America's Craft Breweries. *George Washington Journal of Energy & Environmental Law* 7(2):164–76.

Edwards, David N. 2003. Ancient Egypt in the Sudanese Middle Nile: A Case of Mistaken Identity? In *Ancient Egypt in Africa*, edited by David O'Connor and Andrew Reid, pp. 137–150. London: Institute of Archaeology, University College.

Edwards, David N. 2014. Early Meroitic Pottery and the Creation of an Early Imperial Culture? In *Dans Ein Forscherleben Zwischen den Welten Zum 80. Geburtstag von Steffen Wenig*, edited by A. Lohwasser and P. Wolf, pp. 51–63. Der Antike Sudan Sonderheft. Berlin: Akademie-Verlag

Edwardson, J.R. 1952. Hops—Their Botany, History, Production, and Utilization. *Economic Botany* 6:160–175.

Emerson, Thomas E. 2003. Materializing Cahokia Shamans. *Southeastern Archaeology* 22:135–154.

Epron, L., and F. Daumas. 1939. *Le Tombeau de Ti. Fascicule I. Les Approches de la Chapelle. Mémoires Publiés par les Membres de L'Institut Français d'Archéologie Orientale du Caire 65*. Cairo: IFAO.

Ericson, Jonathon E., and Timothy G. Baugh (eds). 1993. *The American Southwest and Mesoamerica: Systems of Prehistoric Exchange*. New York: Springer Science.

Espinoza, W. 1987. *Historia del Tahuantinsuyo*. Lima: Ediciones Amaru.

Esterick, Penny Van, and Joel Greer. 1985. Beer Consumption and Third World Nutrition. *Food Policy 10*(1):11–13.

Estete, Miguel De. 1924. Relacion que del descubrimiento y conquista del Peru [1534]. Coleccion de libros y documentos referentes a la historia del Perui, segunda serie (Lima) 8: pp. 3–56.

Farabee, W.C. 1908. *Physiological and Medical Observations among the Indians of Southwestern United States and Northern Mexico. Bulletin, Bureau of American Ethnology 34*:28.

Feeney, R.E. 1997. *Polar Journeys: The Role of Food and Nutrition in Early Exploration*. Fairbanks: University of Alaska Press.

Felding, Louise. 2015. The Egtved Girl: Travel, Trade & Alliances in the Bronze Age. *Adoranten* 5–20.

Feldstein, H.S., and S. Poats (eds). 1990. *Working Together: Gender Analysis in Agriculture*. Vol. 1 *Case Studies*. Vol. 2. *Teaching Notes*. West Hartford, CT: Kumarian.

Fischer, R., and J.S. Mshana. 1994. *Inventory of Income-Generating Activities*. Main Report. Tanga Region, Village Development Programme, Tanga, Tanzania.

Fish, Paul, and Susan K. Fish. 1999. Reflections on the Casas Grandes Regional System from the Northwestern Periphery. In *The Casas Grandes World*, edited by Curtis F. Schaafsma and Carroll L. Riley, pp. 27–42. Salt Lake City: University of Utah Press.

Fontein, Joost. 2006. *The Silence of Great Zimbabwe: Contested Landscapes and the Power of Heritage*. London: UCL Press.

Fontein, Joost. 2016. Rain, Power, Sovereignty, and the Materiality of Signs in Southern Zimbabwe. In *Cultural Landscape Heritage in Sub-Saharan Africa*, edited by John Beardsley, pp. 323–372. Washington, DC: Dumbarton Oaks Research Library and Collection.

Forbes, David. 1970. On the Aymara Indians of Bolivia and Peru. *Journal of Ethnological Society of London 2*:193–305.

Frankfort, Henri. 1978. *Kingship and the Gods*. Chicago: University of Chicago Press.

Freeman, Dena. 1997. Minority Groups in Gamo: The Case of Doko Masho. *Report for the Marginalised Minorities of Southern Ethiopia Project*, Addis Ababa.

Frei, Karin Margarita, et al. 2015. Tracing the Dynamic Life Story of a Bronze Age Female. *Scientific Reports 5*(10,431).

Friedman, R.F. 2009. Hierakonpolis Locality HK29A: The Predynastic Ceremonial Center Revisited. *Journal of the American Research Center in Egypt* 45:79–103.

Friedman, Renee F. 2011. Hierakonpolis. In *Before the Pyramids: The Origins of Egyptian Civilization*, edited by E. Teeter, pp. 33–44. Chicago: Oriental Institute Publications 33.

Friedman, Renee F. 2020. Of Mends and Marks: HK6 Mend-a-Thon. *Nekhen News* 32:22–24.

Friedman, Renee, W. Van Neer, and V. Linseele. 2011. The Elite Predynastic Cemetery at Hierakonpolis: 2009–2010 Update. In *Egypt at Its Origins 3: Proceedings of the Third International Conference "Origins of the State. Predynastic and Early Dynastic Egypt," London, 27th July–1st August*, edited by Renee F. Friedman and P.N. Fiske, pp. 157–191. Leuven, Belgium; Paris; Walpole, MA: Peeters.

Fuller, Dorian. 2003. African Crops in Prehistoric South Asia. In *Food, Fuel, and Fields. Progress in African Archaeology*, edited by K. Newmann, N. Butler, and A.S. Kahleber, pp. 239–272. Koln: Africa Praehistorica 15.

Fuller, Dorian. 2007. Non-human Genetics, Agricultural Origins, and Historical Linguistics in South Asia. In *The Evolution of Human Populations in South Asia*, edited by M.D. Petraglia and B. Allchin, pp. 393–443. Dordrecht: Springer.

Fuller, Dorian. 2009. Silence before Sedentism and the Advent of Cash Crops. In *Linguistics, Archaeology and Human Past in South Asia*, edited by T. Toshiba. pp. 147–187. New Delhi: Manohar.

Furst, P.T. 1972. *Flesh of the Gods: The Ritual Uses of Hallucinogens*. New York: Praeger.

Gagnon, Cleste Marie, C. Fred T. Andrus, Jennifer Ida, and Nicholas Richardson. 2015. Local Water Source Variation and Experimental Chicha de Maíz Brewing: Implications for Interpreting Human Hydroxyapatite δ18O Values in the Andes. *Journal of Archaeological Science: Reports* 4:174–181.

Galanty, Heather. 2016a. Saltwater Brewery Creates Edible Six-Pack Rings, Craft Beer.Com. https://www.craftbeer.com/editors-picks/saltwater-brewery-creates-edible-six-pack-rings/.

Galanty, Heather. 2016b. 6 Breweries Going Uber Local with Ingredients, Craft Beer.Com. https://www.craftbeer.com/craft-beer-muses/6-breweries-going-ber-local-ingredients.

Gamboa, E. 2002. Casas Grades Culture, In *Talking Birds, Plumed Serpents, and Painted Women: The Ceramics of Casas Grandes*, edited by J. Strurh, pp. 41–43. Tucson: Tucson Museum of Art.

Garcilaso de la Vega, El Inca. 1966. *Royal Commentaries of the Incas and General History of Peru. Part One*. Harold V. Livermore, trans. Austin: University of Texas Press.

Gardiner, A.F. 1836. *Narrative of a Journey to the Zooloo Country in South Africa*. London: Crofts.

Garfinkel, Y. 1987. Burnt Lime Products and Social Implications in the Pre-pottery Neolithic Villages of the Near East. *Paléorient* 13(1):69–73.

Garnier N., and S.M. Valamoti. 2016. Prehistoric Wine-Making at Dikili Tash (Northern Greece): Integrating Residue Analysis and Archaeobotany. *Journal of Archaeological Science* 74:195–206.

Garrido-Pena, R., and K. Muñoz-López Astilleros. 2000. Visiones sagradas para los líderes. *Complutum* 11:285–300.

Garstang, J. 1907. *The Burial Customs of Ancient Egypt as Illustrated by Tombs of the Middle Kingdom*. London: Archibald Constable.

Geisler, G., B. Keller, and P. Chuzu. 1985. *The Needs of Rural Women in Northern Province: Analysis and Recommendations*. Lusaka: National Commission for Development Planning.

Gelb, I.J. 1965. The Ancient Mesopotamian Ration System. *Journal of Near Eastern Studies* 24:230–243.

Geller, Jeremy. 1993. Bread and Beer in Fourth-Millennium Egypt. *Food and Foodways* 5(3):255–267.

Gelsinger, B.E. 1981. *Icelandic Enterprise: Commerce and Economy in the Middle Ages*. Columbia: University of South Carolina Press.

Gero, Joan M. 1990. Pottery, Power, and . . . Parties! *Archaeology* 43(2):52–56.

Gero, Joan M. 1992. Feasts and Females: Gender Ideology and Political Meals in the Andes. *Norwegian Archaeological Review* 25(1):15–30.

Ghosh, Sushanta, Lovely Rahaman, David Lalvohbika Kaipeng, Dipankar Deb, Nandita Nath, Prosum Tribedi, and Bipin Kumar Sharma. 2016. Community-Wise Evaluation of Rice Beer Prepared by Some Ethnic Tribes of Tripura. *Journal of Ethnic Foods* 3(4):251–256.

Glennon, Robert. 2018. Could Craft Breweries Help Lead the Way in Water Conservation? *Pacific Standard*. https://psmag.com/environment/crafting-solutions-to-water-shortages-in-brewing.

Goldstein, David John, and Robin Christine Coleman. 2004. Schinus Molle (Anacardiaceae) Chicha Production in the Central Andes. *Economic Botany* 58(4):523–529.

Goldstein, David J., Robin C. Coleman Goldstein, and Patrick R. Williams. 2009. You Are What You Drink: A Sociocultural Reconstruction of Pre-Hispanic Fermented Beverage Use at Cerro Baúl, Moquegua, Peru. In *Drink, Power, and Society in the Andes*, edited by Justin Jennings and Brenda J. Bowser. Gainesville: University Press of Florida.

Goldstein, Paul S. 2003. From Stew-Eaters to Maize-Drinkers: The Chicha Economy and the Tiwanaku Expansion. In *The Archaeology and Politics of Food and Feasting in Early States and Empires*, edited by Tamara L. Bray, pp. 143–172. New York: Kluwer Academic/Plenum.

Goldstein, Paul S. 2005. *Andean Diaspora: The Tiwanaku Colonies and the Origins of South American Empire*. Gainesville: University Press of Florida.

González Holguín, Diego. 1989 [1608]. *Vocabulario de la lengua general de todo el Perú llamada lengua qquichua o del inca*. Lima: Universidad Nacional Mayor de San Marcos.

Göransson, H., G. Lemdahl, and B.M. Johansson. 1995. *Alvastra Pile Dwelling: Palaeoethnobotanical Studies*. Theses and Papers in Archaeology, 6. Lund: Lund University Press.

Green, M. 1999. Trading on Inequality: Gender and the Drinks Trade in Southern Tanzania. *Africa 69*:404–425.

Grigg, David. 2004. Wine, Spirits, and Beer: World Patterns of Consumption. *Geography 89*(2):99–110.

Grimstead, Deanna N., Matthew C. Pailes, Katherine A. Dungan, David L. Dettman, Natalia Martínez Tagüeña, and Amy E. Clark. 2013. Identifying the Origin of Southwestern Shell: A Geochemical Application to Mogollon Rim Archaeomolluscs. *American Antiquity 78*(4):640–661.

Griswold, Max, et al. 2018. Alcohol Use and Burden for 195 Countries and Territories, 1900–2016: A Systematic Analysis for the Global Burden of Disease Study 2016. *The Lancet 392*(10,152):1015–1035.

Guaman Poma de Ayala, Felipe. 1987 [1615]. Nueva Crónica y Buen Gobierno. In *Cronicas de America*, Vol. *29*, edited by J.V. Murra, R. Adorno, and J.L. Urioste. Madrid: Historia 16.

Guerra-Doce, E. 2006a. Exploring the Significance of Beaker Pottery through Residue Analysis. *Oxford Journal of Archaeology 25*(3):247–259.

Guerra-Doce, E. 2006b. Propuesta sobre la Functión y el Significado de la Cerámica Campaniforme a la luz de los Análisis de Contenidos. *Trabajos de Prehistoria 63*(1):69–84.

Gutmann, B. 1926. Das Recht der Dschagga, *Arbeiten zur Entwicklungspsychologie 7*:1–733.

Haaland, Gunnar. 1998. Beer, Blood and Mother's Milk: The Symbolic Context of Economic Behaviour in Fur Society. *Sudan Notes and Records, Volume II*: 53–76.

Haaland, Randi. 1987. *Socio-economic Differentiation in the Neolithic Sudan*. BAR International Series 350. Cambridge Monograph in African Archaeology 20. Oxford: BAR.

Haaland, Randi. 1992. Fish, Pots, and Grain: Early and Mid-Holocene Adaptations in the Central Sudan. *The African Archaeological Review 10*:43–64.

Haaland, Randi. 1995. Sedentism, Cultivation, and Plant Domestication in the Holocene Middle Nile Region. *Journal of Field Archaeology 22*:157–174.

Haaland, Randi, 1999. The Puzzle of the Late Emergence of Domesticated Sorghum in the Nile Valley. In *The Prehistory of Food*, edited by C. Gosden and J. Hather, pp. 397–418. London: Routledge.

Haaland, Randi. 2006. Africa and the Near East: Porridge and Bread; Technology and Symbolism Related to Two Food Systems. In *Beitrage zur Europaischen und Africanischen Archaeologie für Manfred Eggert*, edited by H.P. Wotzka, pp. 243–254. Tübingen: Branckre.

Haaland, Randi. 2007. Porridge and Pot, Bread and Oven: Food Ways and Symbolism in Africa and the Near East from the Neolithic to the Present. *Cambridge Archaeological Journal 17*(2):165–182.

Haaland, Randi. 2011. Crops and Culture: Dispersal of African Millets to the Indian Subcontinent and its Cultural Consequences. *Dhaulagiri Journal of Sociology and Anthropology 5*:1–30.

Haaland, Randi. 2012. Changing Food Ways as Indicators of Emerging Complexity in Sudanese Nubia: From Neolithic Agropastoralists to the Meroitic Civilisation. *Azania: Archaeological Research in Africa 47*(3):327–342.

Haaland, Randi. 2017. Kirwan Memorial Lecture: Nile Valley Archaeology and Darfur Ethnography: The Impact of Women on Cultural Evolution. A Personal Reflection. *Sudan and Nubia: 21*:1–14.

Hagen, R., and P.E. Sparboe (eds). 1998. Hans H. Lilienskiold. Trolldom og ugudelighet i 1600-tallets Finnmark. *Ravnetrykk 18*:1–312.

Haggarty, A. 1991. Machrie Moor, Arran: Recent Excavations at Two Stone Circles. *Proceedings of the Society of Antiquaries of Scotland 121*:51–94.

Haggblade, S. 1984. *The Shebeen Queen or Sorghum Beer in Botswana: The Impact of Factory Brews on a Cottage Industry*, PhD. Dissertation. East Lansing: Michigan State University.

Haggblade, S. 1992. The Shebeen Queen and the Evolution of Botswana's Sorghum Beer Industry. In *Liquor and Labor in Southern Africa*, edited by J. Crush and C. Ambler, pp. 395–412. Athens: Ohio University Press.

Haggblade, Steven, and Wilhelm H. Holzapfel. 2004. Industrialization of Africa's Indigenous Beer Brewing. In *Industrialization of Indigenous Fermented Foods*, edited by K.H. Steinkraus, pp. 271–362. New York: Marcel Dekker, Inc.

Halperin, Rhonda, and Judith Olmstead. 1976. To Catch a Feastgiver: Redistribution among the Dorze of Ethiopia. *Africa: Journal of the International African Institute 46*(2):146–165.

Hammond-Tooke, W. D. 1962. *The Bhaca Society: A People of the Transkeian Uplands, South Africa*. Cape Town: Oxford University Press.

Hampton, Tim. 2012. Pilsen (Plzen), In *The Oxford Companion to Beer*, edited by Garrett Oliver, pp. 650–651. Oxford: Oxford University Press.

Hardy, G., and C.H. Richet. 1933. *L'Alimentation Indigene dans les Colonies Françaises*. Paris: Vigot Frères.

Harlan, J.R., and D. Zohary. 1966. Distribution of Wild Wheats and Barley. *Science 153*:1074–1080.

Harner, M. 1973. *The Jivaro: People of the Sacred Waterfalls*. Garden City, NY: Anchor Press.

Harner, M.J. 1973. Introduction. In *Hallucinogens and Shamanism*, edited by M.J. Harner, pp. xi–xv. Oxford: Oxford University Press.

Harris, R. 1975. *Ancient Sippar: A Demographic Study of an Old-Babylonian City (1894–1595 B.C.)*. Leiden: Nederlands Instituut voor het Nabije Oosten.

Hartman, Louis F., and A.L. Oppenheim. 1950. On Beer and Brewing Techniques in Ancient Mesopotamia. *American Oriental Society 10*:1–55.

Hassan, Fekri A. 1988. The Predynastic of Egypt. *Journal of World Prehistory* 2(2):135–184.

Hasson, Rasha N. 2011. Antibacterial Activity of Water and Alcoholic Crude Extract of Flower *Achillea millefolium*. *Rafidain Journal of Science* 22(5):11–20.

Hastorf, Christine A. 1991. Gender, Space, and Food in Prehistory. In *Engendering Archaeology: Women and Prehistory*, edited by Joan M. Gero and Margaret W. Conkey, pp. 132–159. Cambridge: Blackwell.

Hastorf, Christine A. 2003. Andean Luxury Foods: Special Food for the Ancestors, Deities, and the Élite. *Antiquity* 77:545–554.

Hastorf, Christine A., and Sissel Johannessen. 1993. Pre-Hispanic Political Change and Role of Maize in the Central Andes of Peru. *American Anthropologist* 95(1):115–138.

Hayashida, Frances. 2008. Ancient Beer and Modern Brewers: Ethnoarchaeological Observations of Chicha Production in Two Regions of the North Coast of Peru. *Journal of Anthropological Archaeology* 27:161–174.

Hayashida, Frances. 2009. Chicha Histories: Pre-Hispanic Brewing in the Andes and the Use of Ethnographic and Historical Analogues. In *Drink, Power, and Society in the Andes*, edited by Justin Jennings and Brenda J. Bowser, pp. 232–256. Gainesville: University Press of Florida.

Hayashida, Frances. 2019. Crafting Beer Jars for the Inca on the North Coast of Peru. In *Ceramics of the Indigenous Cultures of South America: Studies of Production and Exchange through Compositional Analysis*, edited by Michael D. Glascock, Hector Neff, and Kevin J. Vaughn, pp. 51–54. Albuquerque: University of New Mexico Press.

Hayden, Brian. 1990. Nimrods, Piscators, Pluckers, and Planters: The Emergence of Food Production. *Journal of Anthropological Archaeology* 9(1):31–69.

Hayden, Brian. 2001. Fabulous Feasts: A Prolegomenon to the Importance of Feasting. In *Feasts: Archaeological and Ethnographic Perspectives on Food, Politics, and Power*, edited by Michael Dietler and Brian Hayden, pp. 23–64. Washington, DC: Smithsonian Institution Press.

Hayden, Brian. 2009. The Proof Is in the Pudding. *Current Anthropology* 50(5): 597–601.

Hayden, Brian. 2014. *The Power of Feasts: From Prehistory to the Present.* Cambridge: Cambridge University Press.

Hayden, Brian, Neil Canuel, and Jennifer Shanse. 2013. What Was Brewing in the Natufian? An Archaeological Assessment of Brewing Technology in Epipaleolithic. *Journal of Archaeological Method and Theory* 20:102–150.

Hegmon, M., K. Hays-Gilpin, R.H. McGuire, A.E. Rautman, & S.H. Schlanger. 2008. Changing Perceptions of Regional Interaction in the Prehistoric Southwest. In *The Archaeology of Regional Interaction: Religion, Warfare, and Exchange across the American Southwest and Beyond,* edited by Michelle Hegmon, pp. 1–21. Boulder: University Press of Colorado.

Helbaek H. 1938. Planteavl. *Aarbøger for nordisk oldkynighed og Histtorie* 1938:116–226.

Helbaek, Hans. 1953. Did Man Once Live by Beer Alone? *American Anthropologist* 55(4): 517–519.

Helbaek H.1966. Vendeltime Farming Products at Eketorp on Öland, Sweden. *Acta Archaeologica* 37:216–222.

Helck, Wolfgang. 1975. *Lexikon der Agyptologie, Band 1, A-Ernte,* edited by W. Helck and E. Otto, pp. 789–792. Weisbaden: Otto Harrassowitz.

Helck, W. 1977. *Lexikon der Agyptologie, Band II, Ernefeste Hordjedef,* edited by W. Helck and E. Otto, pp. 586–589. Weisbaden: Otto Harrassowitz.

Hendrickx, S. 2008. Rough Ware as an Element of Symbolism and Craft Specialization at Hierakonpolis' Elite Cemetery HK6. In *Egypt and Its Origins 2: Proceedings of the International Conference Origin of the State, Predynastic and Early Dynastic Egypt, Toulouse (France), 5th-8th September 2005,* edited by B. Midant-Reynes, Y. Tristant, J. Rowland, and S. Hendrickx, pp. 61–85. Leuven: Peeters; Paris: Dudley, MA.

Heun, M., S. Haldorsen, and K. Vollan. 2008. Reassessing Domestication Events in the Near East: Einkorn and *Triticum urartu. Genome* 51:444–451.

Heun, M., R. Schäfer-Pregl, D. Klawan, R. Castagna, M. Accerbi, B. Borghi, and F. Salamini. 1997. Site of Einkorn Wheat Domestication Identified by DNA Fingerprinting. *Science* 278:1312–1314.

Heusinger, T.O. 1856. *Studien über den Ergotismus, insbesondere sein Auftreten im neunzehnten Jahrhundert; aus Anlas einer Epidemie in Oberhessen im Winter 1855/56.* Marburg: Koch.

Hey, G., C. Bell, A.J. Barclay, P. Bradley, J. Mulville, M. Robinson, M. Taylor, and F. Roe. 2016. The Floodplain, Site 1. In *Yarnton: Neolithic and Bronze Age Settlement and Landscape,* edited by G.H. Bell, C.C. Dennis, and M. Robinson, pp. 277–343. Oxford: Oxford University Press.

Hillman, G.C. 2000. Abu Hureyra 1: The Epiphalaeolithic. In *Village on the Euphrates: From Foraging to Farming at Abu Hureyra,* edited by A.M.T. Moore, G.C. Hillman, and A.J. Legge, pp. 327–398. Oxford: Oxford University Press.

Hines, Nickolaus. 2020. How Beer Brewers Are Embracing Sustainability, SevenFifty Daily. https://daily.sevenfifty.com/how-beer-brewers-are-embracing-sustainability/.

Hingh, A de, and C. Bakels. 1996. Palaeobotanical Evidence for Social Differences? The Example of the Early Medieval Domain of Serris-Les-Ruelles, France. Veget Hist. *Archaeobot* 5:117.

Hiwasaki, Lisa. 2000. Ethnic Tourism in Hokkaido and the Shaping of Ainu Identity. *Pacific Affairs* 73(3):393–412.

Hoalst-Pullen, Nancy, Mark W. Patterson, Rebecca Anna Mattord, and Michael D. Vest. 2014. Sustainability Trends in the Regional Craft Beer Industry. *The Geography of Beer: Regions, Environment, and Societies*, edited by Mark Patterson and Nancy Hoalst-Pullen, pp. 109–116. New York: Springer Press.

Homan, Michael M. 2004. Beer and Its Drinkers: An Ancient Near Eastern Love Story. *Near Eastern Archaeology* 67(2):84–95.

Hooke, D. 1998. *The Landscape of Anglo-Saxon England*. Oxford: Basil Blackwell.

Hornsey, Ian. 1999. *Brewing*. Cambridge: Royal Society of Chemistry.

Hornsey, Ian. 2012. *Alcohol and Its Role in the Evolution of Human Society*. Cambridge: Cambridge University Press.

Hough, J.S. 1991. *The Biotechnology of Malting and Brewing*. Cambridge Series in Biotechnology. Cambridge: Cambridge University Press.

Huang, H.T. 2000. *Science and Civilisation in China. Volume 6: Biology and Biological Technology PV: Fermentations and Food Science*. Cambridge: Cambridge University Press.

Hudson, Charles M., ed. 1979. *Black Drink: A Native American Tea*. Athens: University of Georgia Press.

Hunter, M. 1979. *Reaction to Conquest: Effects of Contact with Europeans on the Pondo of South Africa*. Cape Town: David Philip.

Huysecom, E., et al. 2009. The Emergence of Pottery in Africa during the Tenth Millennium cal BC: New Evidence from Ounjougou (Mali). *Antiquity* 83(322):905–917.

Iltis, H.H. 2000. Homeotic Sexual Translocations and the Origin of Maize (Zea mays, Poaceae): A New Look at an Old Problem. *Economic Botany* 54(1):7–42.

Irigoyen-Rascón, Fructuoso. 2015. *Tarahumara Medicine: Ethnobotany and Healing among the Rarámuri of Mexico*. Norman: University of Oklahoma Press.

Isbell, W.H. 1997. *Mummies and Mortuary Monuments: A Postprocessual Prehistory of Central Andean Social Organization*. Austin: University of Texas Press.

Jackson, M.A. 2004. The Chimú Sculptures of Huacas Tacaynamo and El Dragon, Moche Valley, Peru. *Latin American Antiquity* 15(3):298–322.

Jacobsen, Thorkild. 1987. The Graven Image. In *Ancient Israelite Religion: Essays in Honor of Frank Moore Cross*, edited by Patrick Miller Jr., Paul Hanson, and S. Dean, pp. 15–32. Philadelphia: McBride, Fortress.

Jacomet, S. 2007. Neolithic Plant Economies in the Northern Alpine Foreland from 5500–3500 BC cal. In *The Origins and Spread of Domestic Plants in Southwest Asia and Europe*, edited by S. Colledge and J. Conolly, pp. 221–258. Walnut Creek, CA: Left Coast Press.

Janssen, Jozef. 1975. *Commodity Prices from the Ramessid Period: An Economic Study of the Village of Necropolis Workmen at Thebes*. Leiden: Brill Publishers.

Janusek, John Wayne. 2003. Vessels, Time, and Society: Toward a Ceramic Chronology in the Tiwanaku Heartland. In *Tiwanaku and Its Hinterland: Archaeology and Paleoecology of an Andean Civilization, vol. 2, Urban and Rural Archaeology*, edited by Alan Kolata, pp. 30–89. Washington, DC: Smithsonian Institution Press.

Jennings, Justin. 2003. Inca Imperialism, Ritual Change, and Cosmological Continuity in the Cotahuasi Valley of Peru. *Journal of Anthropological Research* 59(4):433–462.

Jennings, Justin. 2005. La Chichera y El Patrón: Chicha and the Energetics of Feasting in the Prehistoric Andes. *Archaeological Papers of the American Anthropological Association* 14:241–259.

Jennings, Justin, Kathleen L. Anrobus, Sam J. Atencio, Erin Glavich, Rebecca Johnson, German Loffler, and Christine Luu. 2005. "Drinking Beer in a Blissful Mood": Alcohol Production, Operational Chains, and Feasting in the Ancient World. *Current Anthropology* 46(2):275–303.

Jennings, Justin, and Brenda J. Bowser. 2009. Drink, Power, and Society in the Andes: An Introduction. In *Drink, Power, and Society in the Andes*, edited by Justin Jennings and Brenda Bowser, pp. 1–27. Gainesville: University Press of Florida.

Jennings, Justin, and Melissa Chatfield. 2009. Pots, Brewers, and Hosts: Women's Power and the Limits of Central Andean Feasting. In *Drink, Power, and Society in the Andes*, edited by Justin Jennings and Brenda J. Bowser, pp. 200–231. Gainesville: University Press of Florida.

Jennings, Justin, and Guy Duke. 2015. Making a Typical Exceptional: The Elevation of Inca Cuisine. In *The Oxford Handbook of the Incas*, edited by Sonia Alconini and Alan Covey, pp. 303–321. Oxford: Oxford University Press.

Jin, G., Y. Zhu, and Y. Xu. 2017. Mystery behind Chinese Liquor Fermentation, *Trends in Food Science and Technology* 63:18–28.

Job, Jayme. n.d. *Foodways and Pottery Use through Time: A Ceramic Use-Alteration Analysis at Neolithic Halai*. PhD Dissertation, SUNY Binghamtom, Binghamtom, NY.

Joffe, Alexander H. 1998. Alcohol and Social Complexity in Ancient Western Asia. *Current Anthropology* 29(3):297–322.

Jones, Ellis. 2018. Brewing Green: Sustainability in the Craft Beer Movement. In *Craft Beverages and Tourism, Volume 2: Environmental, Societal, and Marketing Implications*, edited by Susan L. Slocum, Carol Kline, and Christina T. Cavaliere, pp. 9–26. Cham, Switzerland: Palgrave MacMillan, Springer Nature.

Karp, I. 1980. Beer Drinking and Social Experience in an African Society: An Essay in Formal Sociology. In *Explorations in African Systems of Thought*, edited by I. Karp and C.S. Bird, pp. 83–119. Bloomington: Indiana University Press.

Katz, P.C. 1979. National Patterns of Consumption and Production of Beer. In *Fermented Foods in Nutrition*, edited by C. Gastineau. New York: Academic Press.

Katz, Solomon H., and Mary M. Voigt. 1986. Bread and Beer: The Early Use of Cereals in the Human Diet. *Expedition 28*(2):23–34.

Kaulicke, Peter. 2015. Inka Conceptions of Life, Death, and Ancestor Worship. In *The Inka Empire: A Multidisciplinary Approach*, edited by Izumi Shimada, pp. 247–264. Austin: University of Texas Press.

Keightley, David N. 1978. *Sources of Shang History: The Oracle-Bone Inscriptions of Bronze Age China*. Berkeley: University of California Press.

Keightley, David N. 1989. The Origins of Writing in China: Scripts and Cultural Contexts. In *The Origins of Writing*, edited by Wayne M. Senner, pp. 171–202. Lincoln: University of Nebraska Press.

Keightley, David N. 1994. Sacred Characters. In *Cradles of Civilization: China*, edited by Robert E. Murowchick, pp. 70–79. University of Oklahoma Press, Norman.

Keightley, David N. 1996. Art, Ancestors, and the Origins of Writing in China. *Representations 56*:68–95.

Kemp, B. 1986. *Amarna Reports III. Occasional Publications 4*. London: Egyptian Exploration Society.

Kemp, B. 1989. *Ancient Egypt: Anatomy of a Civilization*. London: Routledge.

Kennedy, John G. 1963. The Role of Beer in Tarahumara Culture. *American Anthropologist 65*(3):620–640.

Kennedy, John G. 1978. *Tarahumara of the Sierra Madre: Beer, Ecology, and Social Organization*. Arlington Heights, IL: AHM Publishing.

Kenoyer, Jonathan Mark. 1998. *Ancient Cities of the Indus Valley Civilization*. Oxford: Oxford University Press.

Kenyon, Kathleen. 1957. *Digging up Jericho*. London: Benn.

King, Daniel J., Michael T. Searcy, Chad L. Yost, and Kyle Waller. 2017. Corn, Beer, and Marine Resources at Casas Grandes, Mexico: An Analysis of Prehistoric Diets Using Microfossils Recovered from Dental Calculus. *Journal of Archaeological Science: Reports 16*:365–379.

King, Malcolm, Alexandra Smith, and Michael Gracey. 2009. Indigenous Health Part 2: The Underlying Causes of the Health Gap. *The Lancet* 374:76–85.

Koch, E., 2003. Mead, Chiefs and Feasts in Later Prehistoric Europe. In *Food, Culture and Identity in the Neolithic and Early Bronze Age*, edited by M.P. Pearson, pp. 125–143. BAR International Series 1117. Oxford: Archaeopress.

Kolar, Miriam A. 2017. Sensing Sonically at Andean Formative Chavín de Huántar, Perú, *Time and Mind* 10(1):39–59.

Kondo, Keiji. 2004. Beer and Health: Preventive Effects of Beer Components on Lifestyle-Related Diseases. *Biofactors* 22:303–310.

Kovacs, Maureen Gallery. 1989. *The Epic of Gilgamesh*. Palo Alto, CA: Stanford University Press.

Kristiansen, K., and B.T. Larsson. 2005. *The Rise of Bronze Age Society, Travels, Transmissions and Transformations*. Cambridge: Cambridge University Press.

Kroll, H. 1981. Mittelneolithisches Getreide aus Dannau. *Offa* 38:85–90.

Kubiak-Martens, L., and J.J. Langer. 2008. Predynastic Beer Brewing as Suggested by Botanical and Physicochemical Evidence from Tell el-Farkha, Eastern Delta. In *Egypt at Its Origins 2*, edited by Béatrix Midant-Reynes, Joanne Rowland, Stan Hendrickx, and Yann Tristant, pp. 427–441. Leuven, Paris, and Dudley, MA: Peeters.

Kuniholm, Peter Ian, and Maryanne W. Newton, with contributions by Richard F. Liebhart. 2011. Dendrochronology at Gordion, In *The New Chronology of Iron Age Gordion*, edited by C. Brian Rose and Gareth Darbyshire, pp. 79–122. Philadelphia: University of Pennsylvania Museum of Archaeology and Anthropology.

La Barre, Weston. 1938. Native American Beers. *American Anthropologist, New Series* 40(2):224–234.

Lagerås, Per. 2000. Burial Rituals Inferred from Palynological Evidence: Results from a Late Neolithic Stone Cist in Southern Sweden. *Journal of Vegetation History and Arcahaeobotany* 9:169–173.

La Hausse, Paul. 1988. *Brewers, Beerhalls and Boycotts: A History of Liquor in South Africa*. Johannesburg: Ravan Press.

Langlois, A. 2016. The Female Tavern-Keeper in Mesopotamia: Some Aspects of Daily Life. In *Women in Antiquity: Real Women across the Ancient World*, edited by S.L. Budin and J.M. Turfa, pp. 113–125. London: Routledge.

Larsson, Mikael, Andreas Svensson, and Jan Apel. 2019. Botanical Evidence of Malt for Beer Production in Fifth-Seventh Century Uppåkra, Sweden. *Archaeological and Anthropological Sciences* 11(5):1961–1972.

Lau, George. 2002. Feasting and Ancestor Veneration at Chinchawas, North Highlands of Ancash, Peru. *Latin American Antiquity* 13(3): 279–304.

Leahy, K. 2007. *"Interrupting the Pots": The Excavation of Cleatham Anglo-Saxon Cemetery*. Council for British Archaeology Research Report 155. Lincolnshire. York.

LeClant, J. 1981. The Empire of Kush: Napata and Meroe. In *General History of Africa II: Ancient Civilizations of Africa*, edited by G. Mokhtar, pp. 278–297. Berkeley: University of California Press.

Lee, Christina. 2007. *Feasting the Dead: Food and Drink in Anglo-Saxon Burial Rituals*. Woodbridge, UK: Boydell Press.

Lerner, Louise. 2018. *Oriental Institute Excavation Finds Large Buildings, Clay Sealings, Evidence of Metallurgy*. University of Chicago News. https://news.uchicago.edu/story/newly-discovered-buildings-reveal-clues-ancient-egyptian-dynasties.

Lev-Yadun, S., A. Gopher, and S. Abbo. 2000. The Cradle of Agriculture. *Science 288*:1602–1603.

Li, Xueqin, Garman Harbottle, Juzhong Zhang, and Changsui Wang. 2002. The Earliest Writing? Sign Use in the Seventh Millennium BC at Jiahu, Henan Province, China. *Antiquity 77*:31–44.

Lieberman, Daniel E., Mickey Mahaffey, Silvino Cubesare Quimare, Nicholas B. Holowaka, Ian J. Wallace, and Aaron L. Baggish. 2020. Running in Tarahumara (Rarámuri) Culture: Persistence Hunting, Footracing, Dancing, Work, and the Fallacy of the Athletic Savage. *Current Anthropology 61*(3):356–379.

Lightfoot, Emma, Tamsin C. O'Connell, Rhiannon E. Stevens, Julie Hamilton, Gill Hey, and Robert E.M. Hedges. 2009. An Investigation into Diet at the Site of Yarnton, Oxfordshire, Using Stable Carbon and Nitrogen Isotopes. *Oxford Journal of Archaeology 28*(3):301–322.

Lindblom, G. 1941. Drinking-Tubes, Especially in Africa. *Ethnos 6*(1–2):48–74.

Linduff, K.M., H. Rubin, and S. Shuyun, eds. 2000. *The Beginnings of Metallurgy in China*. New York: The Edwin Mellen Press.

Liu, Li. 1996. Settlement Patterns, Chiefdom Variability, and the Development of Early States in North China. *Journal of Anthropological Archaeology 15*:237–288.

Liu, Li. 2000. The Development and Decline of Social Complexity in China: Some Environmental and Social Factors. *Indo-Pacific Prehistory Association Bulletin (Maelaka Papers) 20*(4):14–33.

Liu, Li. 2004. *The Chinese Neolithic: Trajectories to Early States*. New York: Cambridge University Press.

Liu, Li, et al. 2019. The Origins of Specialized Pottery and Diverse Alcohol Fermentation Techniques in Early Neolithic China, *PNAS 116*(26):12,767–12,774.

Liu, Li, Jiajing Wang, Danny Rosenberg, Hao Zhoa, György Lengyel, and Dani Nadel. 2018. Fermented Beverage and Food Storage in 13,000 Y-Old Stone

Mortars at Raqefet Cave, Israel: Investigating Natufian Ritual Feasting. *Journal of Archaeological Science: Reports 21*:783–793.

Liu, Li, Jiajing Wang, and Huifang Liu. 2020. *The Brewing Function of the First Amphorae in the Neolithic Yangshao Culture, North China, Archaeologial and Anthropological Sciences 12*(118). DOI: https://doi.org/10.1007/s12 520-020-01069-3.

Logan, Amanda L., Christine A. Hastorf, and Deborah M. Pearsall. 2012. "Let's Drink Together": Early Ceremonial Use of Maize in the Titicaca Basin. *Latin American Antiquity 23*(3):235–258.

Long, Deborah J., Paula Milburn, M. Jane Bunting, Richard Tipping, and Timothy G. Holden. 1999. Black Henbane (*Hyoscyamus niger* L.) in the Scottish Neolithic: A Re-evaluation of Palynological Findings from Grooved Ware Pottery at Balfarg Riding School and Henge, Fife. *Journal of Archaeological Science 26*:45–52.

Lothrop, S.K. 1926. *Pottery of Costa Rica and Nicaragua. Contributions, Museum of the American Indian, Heye Foundation 8.*

Lucas, A., and J.R. Harris. 1962. *Ancient Egyptian Materials and Industries*, 4th ed, revised, J.R. Harris. London: Edward Arnold.

Lumbreras, Luis Guillermo. 1993. *Chavín de Huántar: Excavciones en la Galeria de las Ofrendas*. Mainz Am Rhein, Germany: Verlag Philipp Von Zabern.

Lumbreras, Luis G., and Hernán Amat. 1965. Informe Preliminar sobre las Galerias Interiores de Chavín (primera temporada de trabajos). *Revista del Museo Nacional 34*:143–197.

Lumholtz, D. 1898. *The Huichol Indians of Mexico. Bulletin, American Museum of Natural History 10*(1):11.

Luning, S. 2002. To Drink or Not to Drink: Beer Brewing, Rituals, and Religious Conversions in Maane, Burkina Faso. In *Alcohol in Africa: Mixing Business, Pleasure, and Politics*, edited by D.F. Bryceson, pp. 231–48. Portsmouth, NH: Heinemann.

Luo, M.-C., Z.-L. Yang, F.M. You, T. Kawahara, J.G. Waines, and J. Dvorak. 2007. The Structure of Wild and Domesticated Emmer Wheat Populations, Gene Flow between Them, and the Site of Emmer Domestication. *Theoretical and Applied Genetics 114*:847–959.

Lyumugabe, Francois, Jacques Gros, John Nzungize, Emmanuel Bajyana, and Philippe Thonart. 2012. Characteristics of African Traditional Beers Brewed with Sorghum Malt: A Review. *Biotechnologie, Agronomie, Société et Environnement 16*(4):509–30.

Macdonald, Dave, and Louis Molamu. 1999. From Pleasure to Pain: A Social History of Bsarwa/San Alcohol Use in Botswana. In *Alcohol and Pleasure: A Health Perspective*, edited by S. Peele and M. Grant, pp. 1–7. Philadelphia: Brunner/Mazel.

MacDonell, Anne, ed. 1910 [1669]. *The Closet of Sir Kenelm Digby Knight Opened*. Reprint. London: Philip Lee Warner.

Mackey, Carol. 2010. The Socioeconomic and Ideological Transformation of Farfán under Inka Rule. In *Distant Provinces in the Inka Empire: Toward a Deeper Understanding of Inka Imperialism*, edited by Michael A. Malpass and Sonia Alconini, pp. 221–259. Iowa City: University of Iowa Press.

MacNeish, R.S., A. Garcia Cook, L.G. Lumbreras, R.K. Vierra, and A. Nelken-Terner. 1981. *Prehistory of the Ayacucho Basin, Peru. Vol. 2. Excavations and Chronology*. Ann Arbor: University of Michigan Press.

MacNeish, R.S., R.K. Vierra, A. Nelken-Terner, R. Lurie, and A. Garcia Cook. 1983. *Prehistory of the Ayacucho Basin, Peru. Vol. 4, The Preceramic Way of Life*. Ann Arbor: University of Michigan Press.

Madsen, W., and C. Madsen. 1979. The Cultural Structure of Mexican Drinking Behavior. In *Beliefs, Behaviors, and Alcoholic Beverages*, edited by M. Marshall, pp. 38–54. Ann Arbor: University of Michigan Press.

Mair, L. 1934. *An African People in the Twentieth Century*. London: Routledge & Kegan Paul.

Malville, J. McKim. 2009. Animating the Inanimate: Camay and Astronomical Huacas of Peru. In *Astronomy across Cultures*, edited by J.A. Rubino-Martin, J.A. Belmonte, F. Prada, and A. Alberdi, pp. 261–266. Astronomical Society of the Pacific Conference Series 409.

Malville, J. McKim. 2010. Cosmology in the Inca Empire: Huaca Sanctuaries, State-Supported Pilgrimage, and Astronomy. *Journal of Cosmology* 9:2106–2120.

Mandelbaum, David G. 1965. Alcohol and Culture. *Current Anthropology* 6(3):281–293.

Mangelsdorf, P.C. 1974. *Corn: Its Origin, Evolution, and Improvement*. Cambridge, MA: Harvard University Press.

Mangelsdorf, P.C., R.S. MacNeish, and W.C. Galinat. 1967. Prehistoric Wild and Cultivated Maize. In *The Prehistory of the Tehuacán Valley, Vol. 1, Environment and Subsistence*, edited by D.S. Byers, pp. 178–200. Austin: University of Texas Press.

Manning, Margie. 2018. Recycled Grains from 3 Daughters Brewing Help St. Pete Free Clinic Feed the Hungry. *Catalyst*. https://stpetecatalyst.com/recyc led-grains-from-3-daughters-brewing-help-st-pete-free-clinic-feed-the-hungry/.

Marks, Tasha. 2018. A Sip of History: Ancient Egyptian Beer. *The British Museum Blog*, May 25, 2018. https://blog.britishmuseum.org/a-sip-of-hist ory-ancient-egyptian-beer/.

Marshall, F., & Hildebrand, E. 2002. Cattle before Crops: The Beginnings of Food Production in Africa. *Journal of World Prehistory* 16(2):99–143.

Martin, L., S. Jacomet, and S. Thiébault. 2008. Plant Economy during the Neolithic in a Mountain Context: The Case of "Le Chenet des Pierres" in the

French Alps (Bozel-Savoie, France). *Vegetation History and Archaeobotany* *17*:113–122.

Masters Brewers Association of America. 2020. Diploma and Certificate Programs. https://www.mbaa.com/education/Pages/HEC.aspx.

Matassian, M.K. 1989. *Poisons of the Past. Molds, Epidemics, and History*. New Haven, CT, and London: Yale University Press.

Matthiessen, Peter. 2001. *The Birds of Heaven: Travels with Cranes*. New York: North Point Press.

Mazurowski, Ryszard F., and Bassam Jamous. 2003. Tell Qaramel Excavations 2000. *Polish Archaeology in the Mediterranean 12*:327–341.

McAllister, P.A. 1993. Indigenous Beer in Southern Africa: Functions and Fluctuations. *African Studies 52*(1):71–88.

McCall, Michael K. 2002. Brewers, Woodfuel, and Donors: An Awkward Silence as the Fires Blaze. In *Alcohol in Africa: Mixing Business, Pleasure, and Politics*, edited by Deborah Fahy Bryceson, pp. 93–114. Portsmouth, NH: Heinemann.

McDougall, Christopher. 2009. *Born to Run: A Hidden Tribe, Super Athletes, and Greatest Race the World Has Never Seen*. New York: Knopf Publishing.

McGovern, Patrick E. 2009. *Uncorking the Past: The Quest for Wine, Beer, and Other Alcoholic Beverages*. Berkeley: University of California Press.

McGovern, Patrick E. 2013. A Biomolecular Archaeological Approach to "Nordic Grog." *Danish Journal of Archaeology 2*:112–131.

McGovern, Patrick E., et al. 1999. A Funerary Feast Fit for King Midas. *Nature 402*(23/30 December):863–864.

McGovern, Patrick E., et al. 2005. Chemical Identification and Cultural Implications of a Mixed Fermented Beverage from Late Prehistoric China. *Asian Perspectives 44*(2):249–75.

McGovern, Patrick E, Donald L. Glusker, Lawrence J. Exner, and Mary M. Voigt. 1996. Neolithic Resonated Wine. *Nature 381*(June 6):480–481.

McHugh, James. 2014. Alcohol in Pre-modern South Asia. In *A History of Alcohol and Drugs in Modern South Asia: Intoxicating Affairs*, edited by Harald Fischer-Tiné and Jana Tschurenev, pp. 29–44. New York: Routledge.

Meddens, Frank M., Colin McEwan, and Cirilo Vivanco Pomacanchari. 2010. Inca "Stone Ancestors" in Context at a High-Altitude "Usnu" Platform. *Latin American Antiquity 21*(2):173–194.

Mehdawy, Magda, and Amr Hussein. 2010. *The Pharaoh's Kitchen*. New York: The American University in Cairo Press.

Mekonnen, M.M., and A.Y. Hoekstra. 2010. The Green, Blue and Grey Water Footprint of Crops and Derived Crop Products. Value of Water Research Report Series No.47, UNESCO–IHE.

Mekonnen M.M., and A.Y. Hoekstra. 2011. The Green, Blue and Grey Water Footprint of Crops and Derived Crop Products. *Hydrology and Earth System Sciences 15*(5):1577–1600.

Mendieta, Ramiro Matos, and José Barreiro, eds. 2015. *The Great Inka Road: Engineering an Empire*. Washington, DC: Smithsonian Institution Press.

Meyer, T. 1993. *Wa-konde: maisha, mila na desturi za Wanyakyusa*. Trans. J. Siegrist. Mbeya: Motheco.

Michalowski, Piotr. 1994. The Drinking Gods: Alcohol in Mesopotamian Ritual and Mythology. In *Drinking in Ancient Societies: History and Culture of Drinks in the Ancient Near East*, edited by Lucio Milano, pp. 27–44. Padua: Sargon srl.

Michel, R., P. McGovern, and V. Badler. 1992. Chemical Evidence for Ancient Beer. *Nature 360*(6399).

Michel, Rudolph H., Patrick E. McGovern, and Virginia R. Badler. 1993. The First Wine and Beer: Chemical Detection of Ancient Fermented Beverages. *Analytical Chemistry 65*(8):408–413.

Millard, A. 1988. The Bevelled-Rim Bowls: Their Purpose and Significance. *Iraq 50*:49–57.

Miller, Naomi F., Melinda A. Zeder, and Susan R. Arter. 2009. From Food and Fuel to Farms and Flocks: The Integration of Plant and Animal Remains in the Study of the Agropastoral Economy at Gordion, Turkey. *Current Anthropology 50*(6): 915–924.

Minnis, Paul E. and Michael E. Whalen. 2005. At the Other End of the Puebloan World: Feasting at Casas Grandes, Chihuahua, Mexico. In *Engaged Anthropology: Research Essays on North American Archaeology, Ethnobotany, and Museology*, edited by B. Sunday Eiselt and Michelle Hegmon, pp. 114–128. Museum of Anthropology, Anthropological Papers No. 94. Ann Arbor: University of Michigan Press.

Mittag, Roger. 2014. Geographical Appellations of Beer. In *The Geography of Beer: Regions, Environment, and Societies*, edited by Mark W. Patterson and Nancy Haolst-Pullen, pp. 67–74. New York: Springer Dordrecht Heidelberg.

Mohan, D., and H.K. Sharma. 1995. India. In *International Handbook on Alcohol and Culture*, edited by Dwight B. Heath. pp. 128–141. Westport, CT: Greenwood Press.

Molvik, H. 1992. *Tru og trolldom pa Vestlandet (Beliefs and Witchcraft in Vestlandet)*. Larnes: Vista forlag.

Monckton, H.A. 1966. *A History of English Ale and Beer*. London: The Bodley Head.

Moore, A.M., T.G. Hillman, and A. Legge. 2000. *Village on the Euphrates: From Foraging to Farming at Abu Hureyra*. Oxford: Oxford University Press.

Moore, H.L., and M. Vaughan. 1994. *Cutting down Trees: Gender, Nutrition, and Agricultural Change in the Northern Province of Zambia, 1890–1990*. Portsmouth, NH: Heinemann.

Moore, Jerry D. 1979. Maize Beer in the Economics, Politics, and Religion of the Inca Empire, In *Fermented food beverages in nutrition*. edited by C.F. Gastineau, W.J. Darby, and T.B. Turner, pp. 21–34. New York: Academic Press.

Moore, Jerry D. 1989. Pre-Hispanic Beer in Coastal Peru: Technology and Social Context of Prehistoric Production. *American Anthropologist* 91:682–695.

Moore, Jerry D., and Carol J. Mackey. 2009. The Chimú State. In *Handbook of South American Archaeology*, edited by Helaine Silverman and William H. Isbell, pp. 783–808. New York: Springer Press.

Moroukis, Thomas C. 2010. *The Peyote Road: Religious Freedom and the Native American Church*. Norman: University of Oklahoma Press.

Morris, Craig. 1979. Maize Beer in the Economics, Politics, and Religion of the Inca Empire. In *Fermented Food Beverages in Nutrition*, edited by Clifford F. Gastineau, William J. Darby, and Thomas B. Turner, pp. 21–34. New York: Academic Press.

Morris, Craig, and Donald E. Thompson. 1985. *Huánuco Pampa: An Inca City and Its Hinterland*. New York: Thames and Hudson.

Moseley, Michael E. 1992. *The Incas and Their Ancestors: The Archaeology of Peru*. London: Thames and Hudson.

Moseley, Michael E., Donna J. Nash, Patrick Ryan Williams, Susan D. deFrance, Ana Miranda, and Mario Ruales. 2005. Burning down the Brewery: Establishing and Evacuating an Ancient Imperial Colony at Cerro Baúl, Peru. *Proceedings of the National Academy of Sciences of the United States of American* 102(48):17,264–17,271.

Mukherjee, A.J., A.M. Gibson, and R.P. Evershed. 2008. Trends in Pig Product Processing at British Neolithic Grooved Ware Sites Traced through Organic Residues in Potsherds. *Journal of Archaeological Science* 35:2059–2073.

Mulville, Jaqui, and Mark Robinson, with contributions by Mark Copely, Robert Berstan, Richard Evershed, Fiona Roe, Philippa Bradley, Kate Cramp, Alistair J. Barclay, and Gill Hey. 2016. Food Production and Consumption, In *Yarnton: Neolithic and Bronze Age Settlement and Landscape, Results of Excavations 1990–1998*, edited by Gill Hey, Christopher Bell, Caroline Dennis, and Mark Robinson, pp. 135–154. Thames Valley Landscapes Monograph 39. Oxford: Oxford Archaeology.

Munro, Neil Gordon. 1996. *Ainu Creed and Cult*. London: Routledge.

Murakami, A., P. Darby, B. Javornik, M.S. Pais, E. Seigner, A. Lutz, and P. Svoboda. 2006. Molecular Phylogeny of Wild Hops, *Humulus lupulus* L. *Heredity* 97:66–74.

Murphy, P. 1989. Carbonised Neolithic Plant Remains from The Stumble, An Intertidal Site in Blackwater Estuary, Essex, England. *Circea* 6: 21–38.

Murra, John V. 1960. Rite and Crop in the Inca State. In *Culture in History: Essays in Honor of Paul Radin*, edited by S. Diamond, pp. 393–407. New York: Columbia University Press.

Murra, John V. 1973. Rite and Crop in the Inca State. In *Peoples and Cultures of South America*, edited by Daniel Gross, pp. 377–394. New York: Doubleday/ The Natural History Press.

Murra, John. 1980. *The Economic Organization of the Inca State*. Greenwich, CT: JAI Press.

Myerhoff, Barbara G. 1976. *Peyote Hunt: The Sacred Journey of the Huichol Indians*. Ithaca: Cornell University Press.

Myhre, Knut Christian. 2007. Family Resemblances, Practical Interrelations and Material Extensions: Understanding Sexual Prohibitions, Production and Consumption in Kilimanjaro. *Africa: Journal of the International African Institute* 77(3):207–330.

Narziss, L. 1984. The German Beer Law. *Journal of Institute of Brewing* 90:351–358.

Nash, Donna J. 2010. Fine Dining and Fabulous Atmosphere. In *Inside Ancient Kitchens: New Directions in the Study of Daily Meals and Feasts*, edited by Elizabeth A. Klarich, pp. 83–109. Boulder: University of Colorado Press.

Nash, Donna J. 2012. The Art of Feasting: Building an Empire with Food and Drink. In *Wari: Lords of the Ancient Andes*, edited by Susan E. Gergh, pp. 82–99. Cleveland: Thames and Hudson.

Nash, Donna J. 2019. Craft Production as an Empowering Strategy in an Emerging Empire. *Journal of Anthropological Research* 75(3):238–360.

Nash, Donna J., and Patrick R. Williams. 2009. Wari Political Organization: The Southern Periphery. In *Andean Civilization: A Tribute to Michael E. Moseley*, edited by J. Marcus and P. R. Williams, pp. 257–276. Los Angeles: Cotsen Institute of Archaeology Press, UCLA.

Nash, Donna J., and Susan D. deFrance. 2019. Plotting Abandonment:ExcavatingaRitualDepositattheWariSiteofCerroBaúl.*Journalof Anthropological Archaeology* 53:112–132.

Nelson, Mark L., Andrew Dinardo, Jeffrey Hochberg, and George J. Armlagos. 2010. Brief Communication: Mass Spectroscoptic Characterization of Tetracycline in the Skeletal Remains of an Ancient Population from Sudan, Nubia 350–550 CE. *American Journal of Physical Anthropology* 143:151–154.

Nelson, Max. 2005. *The Barbarian's Beverage: A History of Beer in Ancient Europe*. London: Routledge.

Nelson, Max. 2014. The Geography of Beer in Europe from 1000 BC to AD 1000. In *The Geography of Beer: Regions, Environment, and Societies*, edited by Mark Patterson and Nancy Hoalst-Pullen, pp. 9–22. New York: Springer.

Nesbitt, M., and D. Samuel. 1996. From Staple Crop to Extinction? The Archaeology and History of the Hulled Wheats. In *Hulled Wheat. Proceedings of the First International Workshop on Hulled Wheat*, edited by S. Padulosi, K. Hammer, and J. Heller, pp. 41–100. Rome: International Plant Genetic Resources Institute.

Netherly, Patricia. 1977. *Local Level Lords on the North Coast of Peru.* Ann Arbor, MI: University Microfilms.

Netting, R. 1964. Beer as a Locus of Value among the West African Kofyar. *American Anthropologist* 66:375–384.

Neumann, Hans. 1994. Beer as a Means of Compensation for Work in Mesopotamia during the Ur III Period. In *Drinking in Ancient Societies: History and Culture of Drinks in the Ancient Near East,* edited by Lucio Milano, pp. 321–331. Padua: Sargon srl.

Neve, R.A. 1976. Hops. In *Evolution of Crop Plants,* edited by N.W. Simmonds, pp. 208–212. London: Longman Group.

Nicholson, Edward G. 1960. Chicha Maize Types and Chicha Manufacture in Peru. *Economic Botany* 14(4):290–299.

Nims, Charles F. 1958. The Bread and Beer Problems of the Moscow Mathematical Papyrus. *The Journal of Egyptian Archaeology* 44(1):56–65.

Nissen, Hans. 1986. The Archaic Texts from Uruk. *World Archaeology* 17:317–334.

Nissen, Hans J., Peter Damerow, and Robert K. Eglund. 1993. *Archaic Bookkeeping: Early Writing and Techniques of Economic Administration in the Ancient Near East.* Chicago: University of Chicago Press.

Noble, Gordon, and Kenneth Brophy. 2011. Ritual and Remembrance at a Prehistoric Ceremonial Complex in Central Scotland: Excavations at Forteviot, Perth and Kinross, *Antiquity* 85:787–804.

Noneman, Heidi, Christine VanPool, and Andrew Fernandez. 2017. Examination of Organic Residues and Tribochemical Wear in Low-Fired Casas Grandes Pottery Vessels. Paper Presented at the 82nd Annual Meeting of the Society for American Archaeology, Vancouver, Canada.

Norr, Lynette. 1995. Interpreting Dietary Maize from Bone Stable Isotopes in the American Tropics: The State of the Art. In *Archaeology in the Lowland American Tropics: Current Analytical Methods and Recent Applications,* edited by P.W. Stahl, pp. 198–223. Cambridge: Cambridge University Press.

Novellie, L. 1963. Bantu Beer—Food or Beverage? *Food Industry of South Africa* 16:28.

Nugent, Ruth, and Howard Williams. 2012. Ocular Agency in Early Anglo-Saxon Cremation Burials. In *Encountering Images: Materialities, Perceptions, Relations,* edited by I.M. Back Danielsson, F. Fahlander, and Y. Sjostrand, pp. 187–208. Stockholm Studies in Archaeology, Stockholm University, Stockholm.

Odunfa, S.A. 1985. African Fermented Foods. In *Microbiology of Fermented Foods, Vol. 2,* edited by B.J.B. Wood, pp. 155–199. London and New York: Elsevier Science.

Ogburn, Dennis E. 2004. *Evidence for Long-Distance Transportation of Building Stone in the Inka Empire, from Cuzco, Peru, to Saraguro, Ecuador.* *Latin American Antiquity 15*(4):419–439.

Ohenjo, Nyang'ori, Ruth Willis, Dorothy Jackson, Clive Nettleton, Kenneth Good, and Benon Mugarura. 2006. Health of Indigenous People in Africa. *The Lancet 367*:1937–1946.

Olajire, A.A. 2012. The Brewing Industry and Environmental Challenges. *Journal of Cleaner Production 256*:1–22.

Olalde, Iñigo, et al. 2018. The Beaker Phenomenon and the Genomic Transformation of Northwest Europe. *Nature 555*(7695):190–196.

Orlove, Benjamin, and Ella Schmidt. 1995. Swallowing Their Pride: Indigenous and Industrial Beer in Peru and Bolivia. *Theory and Society 24*:271–298.

Otto, Adelheid. 2014. The Late Bronze Age Pottery of the "Weststadt" of Tell Bazi (North Syria). In *Recent Trends in the Study of Late Bronze Age Ceramics in Syro-Mesopotamia and Neighbouring Regions: Proceedings of the International Workshop in Berlin, 2–5 November 2006/in Collaboration with Claudia Beuger*, edited by Mart Luciani and Arnulf Hausleiter, pp. 85–117. Rahden, Germany: Leidorf.

Oura, E., H. Suomalainen, and R. Viskari. 1982. Breadmaking. In *Economic Microbiology, Vol. 7, Fermented Foods*, edited by A.H. Rose, pp. 88–147, London: Academic Press.

Ouzman, Sven, and Lyn Wadley. 1997. A History in Paint and Stone from Rose Cottage Cave, South Africa. *Antiquity 71*:386–404.

Owen A. (ed. and trans.). 1841. *Ancient Laws and Institutes of Wales, Vol. 1.* London: Eyre and Spottiswoode.

Özdoğan, A. 1999 Çayönü. In *Neolithic in Turkey. Cradle of Civilization*, edited by M. Özdoğan and N. Başgelen, pp. 35–63. Istanbul: Arkeoloji ve Sanat Yayınları.

Özkan, H., A. Brandolini, R. Schafer-Pregl, and F. Salamini. 2002. AFLP Analysis of a Collection of Tretraploid Wheats Indicates the Origin of Emmer and Hard Wheat Domestication in Southeast Turkey. *Molecular Biology and Evolution 19*:1797–1801.

Özkan, H., G. Willcox, A. Graner, F. Salamini, and B. Kilian. 2011. Geographic Distribution and Domestication of Wild Emmer Wheat (*Triticum dicoccoides*). *Genetic Resources and Crop Evolution 58*:11–53.

Ozkaya, V., and A. Coskun. 2009. Kortik Tepe, a New Pre-pottery Neolithic Site in Southeastern Anatolia. *Antiquity 83*(320). http://www.antiquity. ac.uk/projgall/ozkaya320/.

Ozkaya, V., and Oya San. 2007. Körtik Tepe. Bulgular Isığında Kültürel Doku Üzerine Ilk Gözlemler, r. In *Türkiye 'de Neolitik Dönem. Yeni kazılar, yeni bulgular*, edited by M. Özdoğan and N. Başgelen, pp. 21–36. Istanbul: Arkeoloji ve Sanat Yayınları.

Palmer, G.H. 1989. *Cereals in Malting and Brewing*. In *Cereal Science and Technology*, edited by G.H. Palmer, pp. 61–242. Aberdeen: Aberdeen University Press.

Palmer, John, and Colin Kaminski. 2013. *Water: A Comprehensive Guide to Brewers*. Boulder, CO: Brewers Publication.

Parker Pearson, Mike. 1993. The Powerful Dead: Archaeological Relationships between the Living and the Dead. *Cambridge Archaeological Journal* 3(2):203–229.

Parker Pearson, Mike. 2012. *Stonehenge: Exploring the Greatest Stone Age Mystery*. New York: Simon and Schuster.

Parsons, J.J. 1940. Hops in Early California Agriculture. *Agricultural History* 14:110–116.

Paternosto, César. 1996. *The Stone and the Thread: Andean Roots of Abstract Art*. Esther Allen, trans. Austin: University of Texas Press.

Patterson, Mark W., and Nancy Haolst-Pullen. 2014. Geographies of Beer. In *The Geography of Beer: Regions, Environment, and Societies*, edited by Mark W. Patterson and Nancy Haolst-Pullen, pp. 1–5. New York: Springer Dordrecht Heidelberg.

Paulette, Tate. 2020. Archaeological Perspectives on Beer in Mesopotamia: Brewing Ingredients. In *After the Harvest: Storage Practices and Food Processing in Bronze Age Mesopotamia*, edited by Noemi Borelli and Giulia Scazzosi, pp. 65–81. Turnhout, Belgium: Subartu 43, Brepols.

Pearce, H.R. 1976. *The Hop Industry in Australia*. Melbourne: Melbourne University Press.

Pearsall, Deborah M. 2002. Maize Is *Still* Ancient in Prehistoric Ecuador: The View from Real Alto, with Comments on Staller and Thompson. *Journal of Archaeological Science* 29:51–55.

Pearson, Andy. 2017. Refrigeration and Beer. *ASHRAE Journal* 59(6):92.

Perdikaris, S. 1999. From Chiefly Provisioning to Commercial Fishery: Long-Term Economic Change in Arctic, Norway. *World Archaeology* 30:388–402.

Perlov, Diane C. 2009. Working through Daughters: Strategies for Gaining and Maintaining Social Power among the *Chicheras* of Highland Bolivia. In *Drink, Power, and Society in the Andes*, edited by Justin Jennings and Brenda J. Bowser, pp. 49–74, Gainesville, University Press of Florida.

Perrot, J. 1966. Le Gisement Natoufien de Mallaha (Eynan), Israel. *L'Anthropologie* 70:437–483.

Perruchini, E., C. Glatz, M.M. Hald, J. Casana, and J.L. Toney. 2018. Revealing Invisible Brews: A New Approach to the Chemical Identification of Ancient Beer. *Journal of Archaeological Science* 100:176–190.

Perry, Gareth. 2011. Beer, Butter, and Burial: The Pre-burial Origins of Cremation Urns from the Early Anglo-Saxon Cemetery of Cleatham, North Lincolnshire. *Medieval Ceramics* 32:9–21.

Petrie, C.A., and J. Bates. 2017. "Multi-cropping," Intercropping and Adaptation to Variable Environments in Indus South Asia. *Journal of World Prehistory 30*:81–130.

Phillips, Rod. 2014. *Alcohol: A History*. Chapel Hill: University of North Carolina Press.

Phillipson, David. 1994. The Significance and Symbolism of Aksumite Stelae. *Cambridge Archaeological Journal 4*(2):189–210.

Pikirayi, Innocent. 2016. Great Zimbabwe as Power-Scape: How the Past Locates Itself in Contemporary Southern Africa. In *Cultural Landscape Heritage in Sub-Saharan Africa*, edited by John Beardsley, pp. 89–117. Washington, DC: Dumbarton Oaks Research Library and Collection.

Pino Matos, Jose Luis. 2005. El *Ushnu* Inka y La Organización del Espacio en Los Principales *Tampus* de los *Wamani* de la Sierra Central del Chinchaysuyu. *Chungara: Revista de Antropología Chilena 36*(2):303–311.

Piperno, Dolores R. 1988. *Phytolith Analysis*. San Diego: Academic Press.

Piperno, Dolores R. 2003. A Few Kernels Short of a Cob: On the Staller and Thompson Late Entry Scenario for the Introduction of Maize into Northern South America. *Journal of Archaeological Science 30*(7):831–836.

Piperno, Dolores R. 2011. The Origins of Plant Cultivation and Domestication in the New World Tropics: Patterns, Process, and New Developments. *Current Anthropology 52*(4):S453–S470.

Piperno, Dolores R., and Deborah M. Pearsall. 1998. *The Origins of Agriculture in the Lowland Neotropics*. San Diego: Academic Press.

Piperno, Dolores R., Anthony J. Ranere, Irene Holst, Jose Iriarte, and Ruth Dickau. 2009. Starch Grain and Phytolith Evidence of Early Ninth Millennium B.P. Maize from the Central Balsas River Valley, Mexico. *Proceedings of the National Academy of Sciences 106*(13):5019–5024.

Pirazzoli-t'Serstevens, M. 1991. The Art of Dining in the Han Period: Food Vessels from Tomb No. 1 at Mawangdui. *Food and Foodways 4*:209–219.

Pitezel, Todd. 2011. *From Archaeology to Ideology in Northwest Mexico: Cerro de Moctezuma in the Casas Grandes Ritual Landscape*. PhD dissertation, University of Arizona, School of Anthropology, Tucson.

Pizarro, Pedro. 1978 [1572]. *Relacion del descubrimiento y conquista del Peril. Edicion, consideraciones preliminares,Guillermo Lohmann Villena; nota, Pierre Duviols. Pontificia Universidad Catolica del Peru*. Lima: Fondo Editorial.

Platt, B.S. 1955. Some Traditional Alcoholic Beverages and Their Importance in Indigenous African Communities. *Proceedings of the Nutrition Society 14*(2):115–124.

Platt, B.S. 1964. Biological Ennoblement: Improvement of the Nutritive Value of the Foods and Dietary Regimens by Biological Agencies. *Food Technology 18*:662–670.

Poelmans, E., and J.F.M. Swinnen. 2011. A Brief Economic History of Beer. In *The Economics of Beer*, edited by E. Poelmans and J.F.M. Swinnen, pp. 3–28. Oxford: Oxford University Press.

Pollack, Susan. 1999. *Ancient Mesopotamia*. Cambridge: Cambridge University Press.

Pollack, Susan. 2003. Feasts, Funerals, and Fast Food in Early Mesopotamian States. In *The Archaeology and Politics of Food and Feasting in Early States and Empires*, edited by Tamara L. Bray, pp. 17–38. New York: Kluwer Academic/Plenum Publishers.

Pollington, S. 2000. *Leechcraft: Early English Charms, Plantlore, and Healing*. Ely: Anglo-Saxon Books.

Pope, Jeremy W. 2013. Epigraphic Evidence for a "Porridge-and-Pot" Tradition on the Ancient Middle Nile. *Azania: Archaeological Research in Africa* 48(4):473–497.

Postgate, Nicholas, Tao Wang, and Toby Wilkinson. 1995. The Evidence for Early Writing: Utilitarian or Ceremonial. *Antiquity* 69:459–480.

Powell, Jonathan J. 2009. Foreword. In *Beer in Health and Disease Prevention*, edited by Victor R. Preedy. New York: Academic Press.

Powell, Marvin A. 1994. Metron Ariston: Measure as a Tool for Studying Beer in Ancient Mesopotamia. In *Drinking in Ancient Societies: History and Culture of Drinks in the Ancient Near East: Papers of a Symposium held in Rome, May 17–19, 1990*, HANES 6, edited by L. Milano, pp. 91–119. Padua: Sargon.

Prakash, O. 1961. *Food and Drinks in Ancient India*. New Delhi: Munshi Ram Manohar Lal.

Prieto, O. Gabriel. 2011. Chicha Production during the Chimú Period at San José de Moro, Jequetepeque Valley, North Coast of Peru. In *From State to Empire in the Prehistoric Jequetepeque Valley, Peru*, edited by Colleen M. Zori and Llana Johnson, pp. 105–128. Oxford: BAR International Series 2310.

Princeton University, –Office of Communications. 2021. *Archaeological Team Co-led by Princeton's Vischak Identifies World's Oldest Industrial-Scale Brewery at Abydoes, Egypt, ca. 3000 BCE*. Contributors Jamie Saxon and Robert Polner.

Protz, Roger. 2012. Ale-Wives. In *The Oxford Companion to Beer*, edited by Garrett Oliver, pp. 32–33. Oxford: Oxford University Press.

Pryor, Alan. 2009. Indian Pale Ale: An Icon of Empire. Commodities of Empire Working Paper No. 13, pp. 1–21.

Quin, P.J. 1959. *Food and Feeding Habits of the Pedi with Special Reference to Identification, Classification, Preparation and Nutritive Value of Respective Foods*. Johannesburg: Witwatersrand University Press.

Rabin, D., and C. Forget, eds. 1998. *The Dictionary of Beer and Brewing*, 2nd edn. Chicago: Fitzroy Dearborn Publications.

Rakita, Gordon F.M., and Rafael Cruz. 2015. Organization of Production at Paquimé, In *Ancient Paquimé and the Casas Grades World*, edited by Michael E. Whalen and Paul E. Minnis, pp. 58–82. Tucson: University of Arizona Press.

Rahmstorf, L. 2008. The Bell Beaker Phenomenon and the Interaction Spheres of the Early Bronze Age East Mediterranean: Similarities and Differences. In *Constuire le temps. Histoire et méthodes des chronologies et calendriers des derniers millénaires avant notre ère en Europe occidentale*. Actes du XXXe colloque international de Halma-Ipel, UMR 8164 (CNRS, Lille 3, MCC), 7–9 décembre 2006, edited by A. Lehoërff, A., pp. 149–170. Lille, Collection Bibracte 16. Gluxen-Glenne: Bibracte, Centre archéologique européen. Glux-en-Glenne: Bibracte Centre archéologique européen.

Ramirez, Susan Elizabeth.1996. *The World Upside Down: Cross-Cultural Contact and Conflict in Sixteenth Century Peru*. Stanford, CA: Stanford University Press.

Ramirez, Susan Elizabeth.1998. Rich Man, Poor Man, Beggar Man, or Chief: Material Wealth as a Basis of Power in Sixteenth-Century Peru. In *Dead Giveaways: Indigenous Testaments of Colonial Mesoamerica and the Andes*, edited by Susan Kellogg and Matthew Restall, pp. 215–248. Salt Lake City: University of Utah Press.

Reinhard, Johan. 1985. Chavin and Tiahuanaco: A New Look at Two Andean Ceremonial Centers. *National Geographic Research 1* (3): 345–422.

Reinhard, Johan 2007 Machu Picchu: Exploring an Ancient Sacred Center. Los Angeles: Colsten Institute of Archaeology.

Reusch. D. 1998. Imbiza kayibil' ingenambheki: The Social Life of Pots. In *Ubumba: Aspects of Indigenous Ceramics in KWazulu-Natal*, edited by B. Bell and I. Calder, pp. 19–40. Pietermaritzburg: Tatham Art Gallery.

Richards, A.I. 1939. *Land, Labour, and Diet in Northern Rhodesia*. London: Oxford University Press.

Richards, Colin. 1996. Henges and Water: Towards an Elemental Understanding of Monumentality and Landscape in Late Neolithic Britain. *Journal of Material Culture 1*(3):313–336.

Richards, Colin. 2005. The Ceremonial House 2. In *Dwelling among the Monuments: The Neolithic Village of Barnhouse, Maeshowe Passage Grave and Surrounding Monuments at Stenness, Orkney*, edited by C. Richards, pp. 129–156. Cambridge: McDonald Institute for Archaeological Research; Cambridge University Press.

Richards, Colin, et al. 2016. Settlement Duration and Materiality: Formal Chronological Models for the Development of Barnhouse, a Grooved Ware Settlement in Orkney. *Proceedings of the Prehistoric Society 82*:193–225.

Richards, J.D. 1987. The Significance of Form and Decoration of Anglo-Saxon Cremation Urns. Oxford: British Archaeological Reports, British Series 166.

Rick, John W. 2005. The Evolution of Authority and Power at Chavín de Huántar, Peru. *Archaeological Papers of the American Anthropological Association* 14:71–89.

Rick, John W. 2006. Chavín de Huántar: Evidence for an Evolved Shamanism. In *Mesas and Cosmologies in the Central Andes*, pp. 101–112. San Diego Museum Papers 44.

Robbins, R.H. 1979. Problem-Drinking and the Integration of Alcohol in Rural Burganda. In *Beliefs, Behaviors, and Alcoholic Beverages: A Cross-Cultural Survey*, edited by M. Marshall, pp 351–361. Ann Arbor: University of Michigan Press.

Robicsek, F. 1978. *The Smoking Gods: Tobacco in Mayan Art, History, and Religion*. Norman: University of Oklahoma Press.

Robinson, D. 1994. Plants and Vikings: Everyday Life in Viking Age Denmark. *Botanical Journal of Scotland* 46(4):542–551.

Rojo-Guerra, Manuel Ángel, Rafael Garrido-Pena, Íñigo Garcia-Martínez-De-Lagrán, Jordi Juan-Treserras, and Juan Carlos Matalmala. 2006. Beer and Bell Beakers: Drinking Rituals in Copper Age Inner Iberia. *Proceedings of the Prehistoric Society* 72:243–265.

Rollefson, Gary O. 1983. Ritual and Ceremony at Neolithic Ain Ghazal (Jordan). *Paléorient* 9(2):29–83.

Roller, Lynn E. 1983. The Legend of Midas. *Classical Antiquity* 2(2):299–313.

Rösch, Manfred. 2005. Pollen Analysis of the Contents of Excavated Vessels—Direct Archaeobotanical Evidence of Beverages. *Vegetation History and Archaeobotany* 14(3):179–188.

Roscoe, John. 1923. *The Bakitara or Banyoro: The First Part of the Report of the Mackie Ethnological Expedition to Central Africa*. Cambridge: Cambridge University Press.

Rose, C. Brian. 2012. Introduction: The Archaeology of Phrygian Gordion. In *The Archaeology of Phrygian Gordion, Royal City of Midas*, edited by C.B. Rose, pp. 1–19. Philadelphia: University of Pennsylvania Museum.

Rosenberg, Michael, and Richard W. Redding. 2000. Hallan Çemi and Early Village Organization in Eastern Anatolia. In *Live in Neolithic Farming Communities: Social Organization, Identity, and Differentiation*, edited by Ian Kuijt, pp. 39–62. New York: Kluwer Academic/Plenum Publishing.

Rostworowski de Diez Canseco, Maria. 1989. *Costa Peruana Prehispánica*. Lima: Instituto de estudios Peruanos.

Roy, J.K. 1978. Alcoholic Beverages in Tribal India and Their Nutritional Role. *Man in India* 58:263–276.

Salazar, Lucy C. 2004. Machu Picchu: Mysterious Royal Estate in the Cloud Forest. In *Machu Picchu, Unveiling the Mystery of the Incas*, edited by R. Burger and L. Salazar, pp. 21–47. New Haven, CT: Princeton University Press.

Salomon, Frank, and George L. Urioste. 1991. *The Huarochiri Manuscript: A Testament of Ancient and Colonial Andean Religion*. Austin: University of Texas Press.

Sams, Kenneth. 1977. Beer in the City of Midas. *Archaeology 30*(2):108–115.

Samuel, Delwen. 1996a. Investigating Ancient Egyptian Baking and Brewing Methods by Correlative Microscopy. *Science 273*:488–490.

Samuel, Delwen. 1996b. Archaeology of Ancient Egypt Beer. *Journal of the American Society of Brewing Chemist 54*:3–12.

Samuel, Delwen. 2000. Brewing and Baking. In *Ancient Egyptian Materials and Technology*, edited by Paul T. Nicholson and Ian Shaw, pp. 537–576. Cambridge: Cambridge University Press.

Sandars, N.K. 1972. *The Epic of Gilgamesh*. London: Penguin Books.

Sangree, W.H. 1962. The Social Functions of Beer Drinking in Bantu Tiriki. In *Society, Culture, and Drinking Patterns*, edited by D.J. Pittman and C.R. Snyder, pp. 6–21. New York: Wiley.

Sarpaki, A. 2009. Knossos, Crete: Invaders, "Sea-Goers," or Previously "Invisible," the Neolithic Plant Economy Appears Fully-Fledged in 9,000 BP. In *From Foragers to Farmers: Papers in Honour of Gordon C. Hillman*, edited by A. Fairbairn, and E. Weiss, pp. 220–234. Oxford: Oxbow Books.

Saul, M. 1981. Beer, Sorghum, and Women: Production for the Market in Rural Upper Volta. *Africa 51*:746–764.

Sayce, A.H. 1911. *Second Interim Report on the Excavations at Meroe in Ethiopia II. The Historical Results. LAAAA, IV*:53–65.

Schaedel, Richard P. 1988. *La Etnografía muchik de las Fotografías de H. Brüning 1886–1925*. Lima: Ediciones COFIDE (Corporación Financiera de Desarrollo).

Schaefer, Stacy B., and Peter T. Furst, eds. 1996. *People of the Peyote: Huichol Indian History, Religion, and Survival*. Albuquerque: University of New Mexico Press.

Schmidt, Peter R., and Matthew C. Curtis. 2000. Urban Precursors in the Horn: Early 1st-Millennium BC Communities in Eritrea. *Antiquity 75*(290):849–858.

Schultes, Richard Evans. 1938. The Appeal of Peyote (*Lophophora williamsii*) as a Medicine. *American Anthropologist 40*(4):698–715.

Seidl, Conrad. 2012. Rauchbier. In *The Oxford Companion to Beer*, edited by Garrett Oliver, pp. 687–688. Oxford: Oxford University Press.

Seligman, B. Z., and Neil Gordon Munro. 1938. Yaikurekarapa: An Old Ainu Oration. *Man 38*:37–40.

Sewell, Steven L. 2014. The Spatial Diffusion of Beer from its Sumerian Origins to Today. In *The Geography of Beer: Regions, Environment, and Societies*, edited by M. Patterson and N. Hoalst-Pullen. New York: Springer Nature.

Sharma, H.K., and D. Mohan. 1999. Perspectives of Alcohol Consumption in India. In *Alcohol and Pleasure: A Health Perspective*, edited by S. Peele and

M. Grant, pp. 101–112. International Centre for Alcohol Policies, Brunner/ Mazel.

Sharma, H.K., B.M. Tripathi, and Pertti J. Pelto. 2010. The Evolution of Alcohol Use in India. *AIDS Behavior 14*:S8–S17.

Sheinbaum, Hilary. 2018. How the 4 Yuengling Sisters Manage the Family Business. *Forbes*, April 30. https://www.forbes.com/sites/hilarysheinb aum/2018/04/30/how-4-sisters-manage-the-family-business-and-still-get-along-and-you-can-too/#6f1ca2e62ca9.

Sherratt, A. 1987. Cups That Cheered. In *Bell Beakers of the Western Mediterranean*, edited by W. Waldren and R.C. Kennard, pp. 81–114. British Archaeological Record International Series 287. Reprinted in *Economy and Society in Prehistoric Europe*, edited by Andrew Sherratt, pp. 376–402, Princeton, NJ: Princeton University Press.

Sherratt, A. 1991. Sacred and Profane Substances: The Ritual Use of Narcotics in Later Neolithic Europe. In *Sacred and Profane: Proceedings of a Conference on Archaeology, Ritual, and Religion*, edited by P. Garwood, D. Jennings, R. Skeates, and J. Toms, pp. 50–64. Oxford University Committee for Archaeology Monograph 32.

Sherratt, A. 1995. Alcohol and Its Alternatives: Symbol and Substance in Pre-industrial Cultures. In *Consuming Habits: Drugs in History and Anthropology*, edited by J. Goodman, P.E. Lovejoy, and A. Sherratt, pp. 11–46. London: Routledge.

Shimada, Izumi. 1994. *Pampa Grande and the Mochica Culture*. Austin: University of Texas Press.

Shimada, I. 2001. Late Moche Urban Craft Production: A First Approximation. In *Moche Art and Archaeology in Ancient Peru, Studies in the History of Art 61, Center for Advanced Study in the Visual Arts Symposium Papers XL*, edited by J. Pillsbury, pp. 177–206. New Haven, CT: National Gallery of Art, Yale University Press.

Shortland, A.J., and M.S. Tite. 2000. Raw Materials of Glass from Amarna and Implications for the Origin of Egyptian Glass. *Archaeometry 42*:141–151.

Simmons, Alan H., et al. 1990. A Plastered Human Skull from Neolithic 'Ain Ghazal, Jordan. *Journal of Field Archaeology 17*(1):107–110.

Simpson, Elizabeth. 2012. Royal Phrygian Furniture and Fine Wooden Artifacts from Gordion. In *The Archaeology of Phrygian Gordion, Royal City of Midas*, edited by C.B. Rose, pp. 149–164. Philadelphia: University of Pennsylvania Museum.

Singh, G., and B. Lal. 1979. Alcohol in India: A Review of Cultural Traditions and Drinking Practices. *Indian Journal of Psychiatry 21*:39–45.

Sitka, Hans-Peter. 2011a. Early Iron Age and Late Medieval Malt Finds from Germany—Attempts at Reconstruction of Early Celtic Brewing and the Taste of Celtic Beer. *Journal of Archaeological and Anthropological Science 3*:41–48.

Sitka, Hans-Peter. 2011b. Beer in Prehistoric Europe. In *Liquid Bread: Beer and Brewing in Cross-Cultural Perspective*, edited by Wulf Schiefenhovel and Helen MacBeth, pp. 55–62. New York: Berghahn Books.

Sloan, Timothy R. 2012. Green Beer: Incentivizing Sustainability in California's Brewing Industry. *Golden Gate University Environmental Law Journal* 5(2):481–507.

Slotkin, James S. 1956. *The Peyote Religion: A Study in Indian-White Relations*. Glencoe: Free Press.

Smalley, John, and Michael Blake. 2003. Sweet Beginnings Stalk Sugar and the Domestication of Maize. *Current Anthropology* 44(5):675–703.

Smith, Barbara Li, and Yun Kuen Lee. 2008. Mortuary Treatment, Pathology, and Social Relations of the Jiahu Community. *Asian Perspectives* 47(2):242–298.

Smith, Bruce D. 1989. Origins of Agriculture in Eastern North America. *Science* 246(4937):1566–1571.

Smith, K. Annabelle. 2013. Hey Vegans! There May be Fish Bladder in Your Guinness. *Smithsonian Magazine*, March 13.

Sozen, Erol, and Martin O'Neill. 2017. An Exploration of the Motivations Driving New Business Start-up in the United States Craft Brewing Industry. In *Craft Beverages and Tourism: Environmental, Societal, and Marketing Implications*, edited by Susan L. Slocum, Carol Kline, and Christina T. Cavaliere, pp. 195–212. Cham, Switzerland: Palgrave Macmillan, Springer Nature.

Spalinger, Anthony. 1988. Dates in Ancient Egypt. *Studien zur Altägyptischen Kultur* 15:255–276.

Sperber, D. 1975. Paradoxes of Seniority among the Dorze. In *Proceedings of the First U.S. Conference on Ethiopian Studies, 1973*, edited by H.G. Marcus, pp. 209–222. East Lansing: Michigan State University.

Speth, John D. 2015. When Did Human Learn to Boil? *PaleoAnthropology* 54–67. http://paleoanthro.org/media/journal/content/PA20150054.pdf.

Spielmann, Katherine A. 2002. Feasting, Craft Specialization, and the Ritual Mode of Production in Small-Scale Societies. *American Anthropologist* 104(1):195–207.

Staller, J.E. 2003. An Examination of the Paleobotanical and Chronological Evidence for an Early Introduction of Maize (*Zea mays* L.) into South America: A Response to Pearsall. *Journal of Archaeological Science* 30:373–380.

Staller, J.E., and R.G. Thompson. 2002. A Multidisciplinary Approach to Understanding the Initial Introduction of Maize into Coastal Ecuador. *Journal of Archaeological Science* 29:33–50.

Stanish, Charles. 1997. Nonmarket Imperialism in the Prehispanic Americas: The Inca Occupation of the Titicaca Basin. *Latin American Antiquity* 8(3): 195–216.

Stanish, Charles. 2001. Regional Research on the Inca. *Journal of Archaeological Research* 9(3):213–241.

Stanish, Charles, and Brian. S. Bauer, eds. 2004. *Archaeological Research on the Islands of the Sun and the Moon, Lake Titicaca, Bolivia: Final Results from the Proyecto Tiksi Kjarka.* Cotsen Institute of Archaeology, Monograph 52. Los Angeles: Cotsen Institute of Archaeology, University of California.

Steinkraus, K.H. 1994. Nutritional Significance of Fermented Foods. *Food Research International* 27(3): 259–267.

Stevenson, Alice. 2016. The Egyptian Predynastic and State Formation. *Journal of Archaeological Research* 24(4):421–468.

Stevenson, Edward G.J., and Lucie Buffavand. 2018. "Do Our Bodies Know Their Ways?" Villagization, Food Insecurity, and Ill-Being in Ethiopia's Lower Omo Valley. *African Studies Review* 61(1):109–133.

Stewart, R.T. 1974. Paleobotanical Investigation: 1972 Season. In *American Expedition to Idalion*, edited by L. Stager, A. Walker. and G.E. Wright, pp. 123–129. Cyprus: *Bulletin of the American Schools of Oriental Research.*

Stika, H.P. 1996. Traces of a Possible Celtic Brewery in Eberdingen-Hochdorf, Kreis Ludwigsburg, Southwest Germany. *Vegetation History and Archaeobotany* 5:81–88.

Stol, Marten. 1994. Beer in Neo-Babylonian Times. In *Drinking in Ancient Societies: History and Culture of Drinks in Ancient Near East: Papers of a Symposium Held in Rome 1990*, History of the Ancient Near East/Studies VI, edited by Lucio Milano, Padova, pp. 156–183. Italy.

Stordeur, D., and F. Abbés. 2002. Du PPNA au PPNB. *Bulletin de la Société Préhistorique Française* 99:563–595.

Suggs, D.N. 1996. Mosadi Tshwene: The Construction of Gender and the Consumption of Alcohol in Botswana. *American Ethnologist* 23:597–610.

Sutton, J.E.G. 1977. The African Aqualithic. *Antiquity* 51:25–34.

Sveinbjarnaddóttir, B, et al. 2007. The Paleoecology of a High Status Icelandic Farm. *Environmental Archaeology* 12:187–206.

Swenson, Edward R. 2006. Competitive Feasting, Religious Pluralism and Decentralized Power in the Late Moche Period. In *Andean Archaeology III: North and South*, edited by W.H. Isbcll and H. Silverman, pp.112–142. New York: Springer.

Swenson, Edward R. 2007. Local Ideological Strategies and the Politics of Ritual Space in the Chimú Empire. *Archaeological Dialogues* 14(1):61–90.

Swenson, Edward R. 2012a. Moche Ceremonial Architecture as Thirdspace: The Politics of Place-Making in the Ancient Andes. *Journal of Social Archaeology* 12(1):3–28.

Swenson, Edward R. 2012b. Warfare, Gender, and Sacrifice in Jequetepeque, Peru. *Latin American Antiquity* 23(2):167–193.

Takakura, Shinichiro. 1960. *The Ainu of Northern Japan, a Study in Conquest and Acculturation*, translated and annotated by John A. Harrison. Transactions of the American Philosophical Society.

Takamiya, I.H. 2008. Firing Installations and Specialization: A View from Recent Excavations at Hierakonpolis Locality 11C. In *Egypt at Its Origins 2*, edited by B. Midant-Reynes and Y. Tristant, pp. 187–202. Leuven: Peeters.

Taylor, John H. 2010. *Journey through the Afterlife: Ancient Egyptian Book of the Dead*. Cambridge, MA: Harvard University Press.

Taylor, Timothy G., Gary F. Fairchild, Alan W. Hodges, and Thomas J. Stevens. 2014. *Economic Contributions of the Florida Craft Brewing Industry to the Florida Economy*. Sponsored Project Report for the Florida Brewers Guild, University of Florida, Gainesville.

Tello, Julio C. 1960. *Chavín: cultura matriz de la civilización Andina. Plubliccion Antropologica del Archivo "Julio C. Tello" de la Universidad Nacional Mayor de Marcos 2*. Lima: Universidad Nacional de San Marcos.

Thomsen, T. 1929. Egekistfunder fra Egtved, fra den Aeldere Bronze Alder. *Nordisk Fortisdminder* 2:165–214.

Tilley, C. 1996. *An Ethnography of the Neolithic, Early Prehistoric Societies in Southern Scandinavia*. Cambridge: Cambridge University Press.

Toivari-Viitala, Jaana. 2011. Deir el-Medina (Development). In *UCLA Encyclopedia of Egyptology*, edited by Willeke Wendrich, pp. 1–15. Los Angeles: UCLA.

Tonsmeire, Michael. 2014. *American Sour Beers: Innovative Techniques for Mixed Fermentations*. Boulder, CO: Brewers Association.

Torres, Constantino Manuel. 2008. Chavín's Psychoactive Pharmacopoeia: The Econographic Evidence. In *Chavín: Art, Architecture, and Culture*, edited by William Conklin and Jeffrey Quilter, pp. 239–259. Monograph 61. Los Angeles: Cotsen Institute of Archaeology, UCLA.

Tresset, Anne, and Jean-Denis Vigne. 2011. Last Hunter-Gatherer and First Farmers of Europe. *Camptes Rendus Biologies* 334:182–189.

Trigger, Bruce G. 1969. The Myth of Meroe and the African Iron Age. *African Historical Studies* 2(1):23–50.

Tschurenev, Jana, and Harald Fischer-Tiné. 2014. Introduction: Indian Anomalies? Drink and Drugs in the Land of Gandhi. In *A History of Alcohol and Drugs in Modern South Asia: Intoxicating Affairs*, edited by Harald Fischer-Tiné and Jana Tschurenev, pp. 1–26. New York: Routledge.

Tylecote, R.F. 1970. *Iron Working at Meroe, Sudan. BHM* 4:67–72.

Tzedakis, Y., H. Martlew, and M.K. Jones. 2008. *Archaeology Meets Science: Biomolecular Investigations in Bronze Age Greece*. Oxford: Oxbow Books.

Underhill, A. 1989. Warfare during the Chinese Neolithic Period: A Review of the Evidence. In *Conflict: Current Archaeological Perspectives*, edited by D.C.

Tkaczuk and B.C. Vivian, pp, 229–237. Calgary: Archaeological Association of the University of Calgary.

Underhill, A. 1994. Variation in Settlements during the Longshan Period of Northern China. *Asian Perspectives 33*(2):197–228.

Underhill, A. 2000. An Analysis of Mortuary Ritual at the Dawenkou Site, Shandong, China. *Journal of East Asian Archaeology 2*(1–2):93–128.

Unger, R. 2001. *A History of Brewing in Holland, 900–1900, Economy, Technology, and the State*. Leiden and Boston: Brill Academic Publishers.

United Nations Department of Economic and Social Affairs. 2021. Indigenous Peoples. https://www.un.org/development/desa/indigenouspeoples/about-us.html#:~:text=Indigenous%20peoples%20are%20inheritors%20and,societies%20in%20which%20they%20live.

Uzendoski, Michael A. 2004. Manioc Beer and Meat: Value, Reproduction, and Cosmic Substance among the Napo Runa of the Ecuadorian Amazon. *The Journal of Royal Anthropological Institute 10*(4):883–902.

Uzendoski, Michael A. 2005. *The Napo Runa of Amazonia Ecuador*. Urbana and Chicago: University of Illinois Press.

Valamoti, Soultana Maria. 2018. Brewing Beer in Wine Country? First Archaeobotanical Indications for Beer Making in Early and Middle Bronze Age Greece. *Vegetation History and Archaeobotany 27*(4):611–625.

Valdez, Lidio M. 2006. Maize Beer Production in Middle Horizon Peru. *Journal of Archaeological Research 62*:53–80.

Valdez, Lidio M. 2012. Molle Beer in a Peruvian Central Highland Valley. *Journal of Anthropological Research 68*(1):71–93.

Vallee, Bert L. 1998. Alcohol in the Western World. *Scientific American 278*(6):80–85.

Vance, Georgia H. 2020. Goddess in the Sheets, Prostitute in the Streets: Examining Public and Private Divisions of Gender in Mesopotamian Cities. *Inquiries Journal 12*(11). http://www.inquiriesjournal.com/a?id=1837.

Vander Linden, M. 2006. Le Phénomène Campaniforme dans l'Europe du 3ème Millénaire avant notre ère. Synthèse et nouvelles perspectives. British Archaeological Reports International Series 1470. Oxford: Archaeopress.

Van Dijk, R. 2002. Modernity's Limits: Pentecostalism and the Moral Rejection of Alcohol in Malawi, In *Alcohol in Africa: Mixing Business, Pleasure and Politics*, edited by D. Bryceson, pp. 249–264. Portsmouth, NH: Heinemann.

van Heerden, I.V. 1987. Nutrient Content of Sorghum Beer Strainings. *South African Journal of Animal Science 17*(4):171–175.

VanPool, Christine S. 2003. The Shaman-Priest of the Casas Grandes Region, Chihuahua, Mexico. *American Antiquity 68*(4):696–717.

VanPool, Todd L., Christine S. VanPool, and R. D. Leonard. 2005. The Casas Grandes Core and Periphery. In *Archaeology Between Borders: Papers from the 13th Biennial Jornada Mogollon Conference*, edited by M. Thompson,

J. Jurgena, and L. Jackson, pp. 25–35. El Paso, TX: El Paso Museum of Archaeology.

VanPool, Christine, MacLaren Law-de-Lauriston, Heidi Noneman, and Andrew Fernandez. 2016. *Booze or Food? Experimental Archaeology of Low-Fired Pottery to Examine Tribochemical Processes.* In Poster Presented at the 81st Annual Meeting of the Society for American Archaeology, Orlando.

VanPool, Todd L., and Christine S. VanPool. 2019. Paquime's Appeal: The Creation of an Elite Pilgrimage Site in the North American Southwest. In *Cognitive Archaeology: Mind, Ethnography, and the Past in South Africa and Beyond*, edited by David Whitley, Johannes Lubser, and Gavin Whitelaw. New York: Routledge.

Viklund, K. 1998. Cereals, Weeds and Crop Processing in Iron Age Sweden. Archaeology and Environment 14. University of Umeå, Department of Archaeology, Environmental Archaeology Laboratory.

Viner, S., J. Evans, U. Albarella, and M. Parker Pearson. 2010. Cattle Mobility in Prehistoric Britain: Strontium Isotope Analysis of Cattle Teeth from Durrington Walls (Wiltshire, Britain). *Journal of Archaeological Science* 37:2812–2820.

Wadley, Lyn. 1989. Legacies from the Later Stone Age. *South African Archaeological Society Goodwin Series* 6:42–53.

Wainwright, G.J., and I.H. Longworth. 1971. *Durrington Walls: Excavations 1966–1968.* London: Society of Antiquaries.

Wang, Jiajing, Renee Friedman, and Masahiro Baba. In press. Predynastic Beer Production, Distribution, and Consumption at Hierakonpolis, Egypt. *Journal of Anthropological Archaeology.*

Wang, Jiajing, Li Liu, Terry Ball, Linjie Yu, Yuanqing Li, and Fulai Xing. 2016. Revealing a 5,000-Y-Old Beer Recipe in China. *Proceeding of the National Academy of Sciences* 113(23):6444–6448.

Wang, Jiajing, Li Liu, Andreea Georgescu, Vivienne V. Le, Madeleine H. Ota, Silu Tang, and Mahpiya Vanderbilt. 2017. Identifying Ancient Beer Brewing through Starch Analysis: A Methodology. *Journal of Archaeological Science: Reports* 15:150–160.

Washburn, Dorothy K., William N. Washburn, and Petia A. Shipkovac. 2011. The Prehistoric Drug Trade: Widespread Consumption of Cacao in Ancestral Pueblo and Hohokam Communities in the American Southwest. *Journal of Archaeological Science* 38:1634–1640.

Watson, Elizabeth E. 1998. *Ground Truths: Power and Land in Konso, Ethiopia.* PhD thesis, Cambridge University.

Watson, Elizabeth E. 2009. *Living Terraces in Ethiopia: Konso Landscape, Culture, and Development.* Woodbridge, UK: James Currey.

Weismantel, M.J. 1991. Maize Beer and Andean Social Transformations: Drunken Indians, Bread Babies, and Chosen Women. *French Issue: Cultural Representation of Food* 106(4):861–879. Baltimore: John Hopkins University Press.

Wente, Edward F. 1961. A Letter of Complaint to the Vizier. *Journal of Near Eastern Studies 20*(4):252–257.

Westbrook, Raymond. 1995. Social Justice in the Ancient Near East. In *Social Justice in the Ancient World*, edited by K.D. Irani and Morris Silver, pp. 149–163. Westport, CT: Greenwood Press.

Whalen, Michael E., and Paul E. Minnis. 2009. *The Neighbors of Casas Grandes: Excavating Medio Period Communities of Northwest Chihuahua, Mexico*. Tucson: University of Arizona Press.

White, Chris, and Jamil Zainasheff. 2010. *Yeast: The Practical Guide to Beer Fermentation*. Boulder, CO: Brewers Publication.

White, Nancy. 2017. *Long-Distance Connections across the Southeastern US and Mesoamerica*. Society for American Archaeology Annual Meeting, Vancouver, Canada.

WHO (World Health Organization). 2017. Diarrhoeal Disease. https://www.who.int/news-room/fact-sheets/detail/diarrhoeal-disease.

WHO (World Health Organization). 2020. The Top 10 Causes of Death. https://www.who.int/news-room/fact-sheets/detail/the-top-10-causes-of-death

Wickham-Jones, Carolina Rosa. 1990. Rhum: Mesolithic and Later Sites at Kinlock, Excavations 1984–86. Society of Antiquaries of Scotland, Monograph Series No. 7, Edinburgh, Scotland.

Wiessner, Pauline. 1996. Introduction: Food, Status, Culture, and Nature. In *Food and the Status Quest: An Interdisciplinary Perspective*, edited by Pauline Wiessner and Wolf Schiefenhoval, pp. 1–18. Providence, RI: Berghahn Books.

Wiessner, Pauline. 2001. Of Feasting and Value: Enga Feasts in a Historical Perspective (Papua New Guinea). In *Feasts: Archaeological and Ethnographic Perspectives on Food, Politics, and Power*, edited by Michael Dietler and Brian Hayden, pp. 115–143. Washington, D.C.: Smithsonian Institution Press.

Wilbert, J. 1987. *Tobacco and Shamanism in South America*. New Haven, CT: Yale University Press.

Wilke, Detlef, and Ina Wunn. 2019. A Multi Theory Approach for Studying Presumed Ritual Pottery—The Stirrup Spout Bottle as a Case Study from the Pre-Hispanic Andes. *Journal of Archaeology and Anthropology 1*(5):1–21.

Willcox, George. 2002. Charred Plant Remains from a 10th Millennium BP Kitchen at Jerf el Ahmar (Syria). *Vegetation History and Archaeobotany 11*:55–60.

Willcox, George, and Danielle Stordeur. 2012. Large-Scale Cereal Processing before Domestication during the Tenth Millennium Cal BC in Northern Syria. *Antiquity 86*:99–114.

Williams, Patrick Ryan, and Donna J. Nash. 2006. Sighting the *Apu*: A GIS Analysis of Wari Imperialism and the Worship of Mountain Peaks. *World Archaeology 38*(3):455–468.

Williams, Patrick Ryan, Donna J. Nash, Joshua M. Henken, and Ruth Ann Armitage. 2019. Archaeometric Approaches to Defining Sustainable Governance: Wari Brewing Traditions and the Building of Political Relationships in Ancient Peru. *Sustainability* 11(2333):1–16.

Willis, Justin. 2002. For Women and Children: An Economic History of Brewing among the Nyakyusa of Southwestern Tanzania. In *Alcohol in Africa: Mixing Business, Pleasure, and Politics*, edited by Deborah Fahy Bryceson, pp. 56–73. Portsmouth, NH: Heinemann.

Wilson, Gay. 1975. Plant Remains from the Graveney Boat and Early History of *Humulus lupulus* L. in W. Europe. *New Phytologist* 75:627–648.

Wilson, Monica. 1954. Nyakyusa Ritual and Symbolism. *American Anthropologist* 56(2): 228–241.

Winter, J.C., ed. 2000. *Tobacco Use by Native Americans: Sacred Smoke and Silent Killer*. Norman: University of Oklahoma Press.

Winterman, Denise. 2014. Beer: The Women Taking over the World of Brewing. *BBC Magazine*, January 23. https://www.bbc.com/news/magazine-25656701.

Wood, Brian J.B. 1994. Technology Transfer and Indigenous Fermented Foods. *Food Research International* 27:269–280.

Wood, Brian J.B., and M.M. Hodge. 1985. Yeast–Lactic Acid Bacteria Interaction and Their Contribution to Fermented Foodstuffs. In *Microbiology of Fermented Foods, Volume 1*, edited by B.J.B. Wood, pp. 263–293. London: Elsevier.

Wooley, C. Leonard. 1934. Ur Excavations II: The Royal Cemetery. London: The British Museum and the University Museum, University of Pennsylvania.

Worley, Fay, et al. 2019. Understanding Middle Neolithic Food and Farming in and around the Stonehenge World Heritage Site: An Integrated Approach. *Journal of Archaeological Science: Reports* 26:1–19.

Wright, E., S. Viner-Daniels, M. Parker Pearson, and U. Albarella. 2014. Age and Season of Pig Slaughter at Late Neolithic Durrington Walls (Wiltshire, UK) as Detected through a New System for Recording Tooth Wear. *Journal of Archaeological Science* 52:497–514.

Wright, Henry T., and Gregory Johnson. 1975. Population, Exchange, and Early State Formation in Southwestern Iran. *American Anthropologist* 77:267–289.

Wright, Katherine I. 1994. Ground-Stone Tools and Hunter-Gatherer Subsistence in Southwest Asia: Implications for the Transition to Farming. *American Antiquity* 59:238–263.

Wright, R.E., and Valencia Zegarra, A. 2004. *The Machu Picchu Guidebook*. Boulder, CO: Johnson Books.

Wylie, Alison. 1989. Interpretive Dilemma. In *Critical Traditions in Contemporary Archaeology*, edited by V. Pinsky and A. Wylie, pp. 18–27. Cambridge: Cambridge University Press.

Yartah, T. 2004. Tell 'Abr 3, Un Village du Néolithique Précéramique (PPNA) sur le Moyen Euphrate. *Première Approche, Paléorient 30–2*:141–158.

Yates, James A. Fellows, et al. 2021. The Evolution and Changing Ecology of the African Hominid Oral Microbiome. *Proceedings of the National Academy of Sciences 118*(20), e2021655118. DOI: 10.1073/pnas.2021655118.

Yaya, Isabel. 2015. Sovereign Bodies: Ancestor Cult and State Legitimacy among the Incas. *History and Anthropology 26*(5):639–660.

Yeshurun, Reuven, Guy Bar-Oz, and Dani Nadel. 2013. The Social Role of Food in the Natufian Cemetery of Raqefet Cave, Mount Carmel, Israel. *Journal of Anthropological Archaeology 32*:511–526.

Yool, Stephen, and Andrew Comrie. 2014. A Taste of Place: Environmental Geographies of the Classic Beer Styles. In *The Geography of Beer: Regions, Environment, and Societies*, edited by Mark Patterson and Nancy Hoalst-Pullen, pp. 99–108. New York: Springer.

Yuengling Brewery. 2015. Sustainability. https://www.yuengling.com/our-brewery/#tab_sustain.

Yuengling Brewery. 2018. Committed to Sustainability for Future Generations. https://www.yuengling.com/and-daughters-blog/committed-to-sustainability-for-future-generations/.

Zammit, I.V. 1979. *The Nutritive Value of Sorghum Beer*. Pretoria: National Food Research Institute.

Zarnkow, M., A. Otto, and B. Einwag. 2011. Interdisciplinary Investigations into the Brewing Technology of the Ancient Near East and the Potential of the Cold Mashing Process. In *Liquid Bread: Beer and Brewing in Cross-Cultural Perspectives*, edited by W. Schiefenhövel and H. Macbeth, pp. 47–54. New York: Berghahn.

Zawaski, Michael J., and J. McKim Malville. 2007. An Archaeoastronomical Survey of Major Inca Sites in Peru. *The Journal of Astronomy in Culture XXI*:20–38.

Zhang, Juzhong, Xinghua Xiao, and Yun Kuen Lee. 2004. The Early Development of Music. Analysis of the Jiahu Bone Flutes. *Antiquity 92*(364): 769–778.

Zhang, Juzhong, and L. Wanli. 2010. Research on Fermented Beverage Discovered in Jaihu. In *Wine in Chinese Culture: Historical, Literary, Social and Global Perspectives*, edited by Peter Kupfer, pp. 75–78. Munster: Verlag.

Zohary, Daniel, Maria Hopf, and Ehud Weiss. 2012. *Domestication of Plants in the Old World*, 4th ed. Oxford: Oxford University Press.

Zori, D., J. Byock, E. Erlcndsson, S. Martin, T. Wake, and K.J. Edwards. 2013. Feasting in Viking Age Iceland: Sustaining a Chiefly Political Economy in a Marginal Environment. *Antiquity, 87*:150–165.

Zuidema, R.T. 1981. Inca Observations of the Solar and Lunar Passages through Zenith and Antizenith at Cusco. In *Archaeoastronomy in the Americas*, edited by R. Williamson, pp. 319–342. Los Altos: Ballena Press.

Zuidema, R.T. 2008. The Inca and His Curacas: Royal Polygyny and the Construction of Power. *Bulletin de l'Institut Francais d'Estudes Andines* *37*(1):47–55.

Zuidema, R.T. 2014. The Ushnus of Cusco and Sacred Centres in Andean Ethnography, Ethnohistory, and Archaeology. In *Inca Sacred Space: Landscape, Site, and Symbol in the Andes*, edited by Frank Meddens, Katie Willis, Colin McEwan, and Nicholas Branch, pp. 5–28. London: Archetype Publications.

Index

For the benefit of digital users, indexed terms that span two pages (e.g., 52–53) may, on occasion, appear on only one of those pages.

Tables and figures are indicated by *t* and *f* following the page number.

6/22